Sketching in Human-Computer Interaction

Makayla Lewis • Miriam Sturdee

Sketching in Human-Computer Interaction

A Practical Guide to Sketching Theory and Application

 Springer

Makayla Lewis
School of Computer Science and
Mathematics
Kingston University
London, UK

Miriam Sturdee
School of Computer Science
University of St Andrews
St Andrews, UK

ISBN 978-3-031-50135-7 ISBN 978-3-031-50136-4 (eBook)
https://doi.org/10.1007/978-3-031-50136-4

© The Editor(s) (if applicable) and The Author(s), under exclusive license to Springer Nature Switzerland AG 2024

This work is subject to copyright. All rights are solely and exclusively licensed by the Publisher, whether the whole or part of the material is concerned, specifically the rights of translation, reprinting, reuse of illustrations, recitation, broadcasting, reproduction on microfilms or in any other physical way, and transmission or information storage and retrieval, electronic adaptation, computer software, or by similar or dissimilar methodology now known or hereafter developed.
The use of general descriptive names, registered names, trademarks, service marks, etc. in this publication does not imply, even in the absence of a specific statement, that such names are exempt from the relevant protective laws and regulations and therefore free for general use.
The publisher, the authors and the editors are safe to assume that the advice and information in this book are believed to be true and accurate at the date of publication. Neither the publisher nor the authors or the editors give a warranty, expressed or implied, with respect to the material contained herein or for any errors or omissions that may have been made. The publisher remains neutral with regard to jurisdictional claims in published maps and institutional affiliations.

This Springer imprint is published by the registered company Springer Nature Switzerland AG
The registered company address is: Gewerbestrasse 11, 6330 Cham, Switzerland

If disposing of this product, please recycle the paper.

Foreword to Sketching in Human-Computer Interaction

Sketches have been an important part of interaction design since the earliest days of digital computing. Initially this may have been because of a lack of computer-based prototyping, indeed even code was usually initially sketched using flow diagrams. However, the benefits of low-fidelity prototyping also became evident early, with paper-based methods as well as tools, notably DENIM by Newman et al. (a prototypical website design tool that supports sketch-input), that emulated this. Sometimes sketches are because you do not know sufficient detail, but they are also because you often do not want too much detail. As Buxton and others have argued, the very incompleteness of a sketch cries out to our imagination, we can fill in the gaps as we please, or simply accept the vagueness … which is itself a form of abstraction.

Sketches are of course used extensively to draw the contents of displays of digital devices, but they equally allow one to draw the device itself, or the context in which it is used. This has been an issue that has concerned me for more than 20 years, encouraging students to sketch not just inwards towards the screen, outwards to encompass the direct users, other people around, and whether they are sitting or standing, indoors or outdoors, in rain or sunshine. More recently it became clear that most of the tools used by professional user-experience designers focus solely on the screen content, and even product-design students jump straight into wireframes, without thinking about blue foam or even CAD (Computer Aided Design) models. Sketching is needed more than ever.

Of course, one of the barriers to sketching, and why many reach instantly for wireframe tools, is fear—embarrassing memories of school art lessons, where one learnt that there were those who were artistic, visited by the muses at birth, and those, the majority of us, who were passed over. Try telling that to the pre-school five-year-old given a crayon or pencil, or even those in nineteenth century where line sketching and painting were seen as one of the normal accomplishments of an educated person.

It is thus wonderful to see this book, addressing this myth of the art-less, and reminding us, as our five-year-old selves knew, that we can all draw and sketch. Makayla and Miriam are perfectly placed to do this, not so much because of their own sketching abilities but because of the way they have been training others, particularly in their popular CHI (The ACM Conference on Human Factors in Computing Systems) courses.

The chapters take you step-by-step from the basic elements of sketches to more complete illustrations and issues including accessibility. As befits the topic, the book is beautifully illustrated throughout; indeed, it is worth having simply to look at. However, I hope you do more than look, and work through the chapters, not to create da Vinci masterpieces but to create your own sketches that communicate, engage and excite.

October 2023   Alan Dix

Acknowledgements

With thanks and love to Paul, my parents Tony and Sarah (who let me sketch on the walls at home!) and my aunt Philippa for endless support, my wonderful colleagues, collaborators, friends, and the ever present Ren and Tiny Cat. Finally Sketching in HCI is a reality—Thanks for the journey, Makayla!—Miriam

With thanks and love to my Mum, Jim, colleagues, collaborators and the wonderful sketching and sketchnotes community—including Mauro, Claire, Ruth, Bruno, Prof. Clayton, Andrea, Claudio, Nicole, Lorraine, Mario, Mike, Rob, James, Shaun and Julia, among others. I am also deeply grateful for the companionship of my many little animal friends (guinea pigs, rabbits and chinchillas) and, of course, Miriam!—Makayla

Contents

1	**Introduction**.	1
	1.1 Welcome!.	1
	1.2 What *Is* a Sketch?.	2
	1.3 Why Use Sketching in HCI?.	3
	1.4 The Basics.	3
	1.4.1 Ideation.	3
	1.4.2 Iteration and Reflection.	4
	1.4.3 Evidence and Documentation.	4
	1.4.4 Communication and Dialogue.	4
	1.5 The Complex.	9
	1.5.1 Elaboration.	10
	1.5.2 Input.	10
	1.5.3 Output.	10
	1.5.4 Tool.	10
	1.6 A Short History of Sketching in HCI.	11
	1.7 How to Use This Book.	11
	1.8 Materials.	12
	1.9 Keeping a Sketchbook.	14
	1.10 Practical Application Tips.	16
	1.11 Hands-On Activities.	17
	1.12 References and Resources.	19
	1.13 A Fun Hands-On Activity for Reflection.	19
	References.	22
2	**The Humble Line**.	25
	2.1 Introduction.	25
	2.2 Forget Everything You Learned at School.	27
	2.3 Hands-On Activities.	28
	References.	35

3	**Seeing the World in Icons**		37
	3.1	Introduction	37
	3.2	Basic Visual Icons	38
	3.3	When and How to Use Detail	39
	3.4	Combined Visual Icons	41
	3.5	Domain Visual Icons	41
	3.6	Hands-On Activities	44
		3.6.1 Visual Icon Grid Activity Sheet	48
	References		51
4	**Text, Connections, and Colour**		53
	4.1	Describing Your Visual World	53
	4.2	Using Text in Sketches	54
	4.3	Simple Connections	60
	4.4	Colour Your World	62
	4.5	Hands-On Activities	63
	References		68
5	**People, Faces, and Actions**		71
	5.1	Beyond the Stick Person	71
	5.2	The Simple Figure	72
	5.3	About Face	73
	5.4	A Handy Guide	75
	5.5	Action Stations!	79
	5.6	Hands-On Activities	81
	References		88
6	**Exploring Visual Narratives**		91
	6.1	What Do We Mean by "Visual Narrative"?	91
	6.2	Vignettes	92
	6.3	Sketchnotes	92
		6.3.1 What?	92
		6.3.2 Why?	95
		6.3.3 Where?	96
		6.3.4 How?	101
	6.4	Comics	105
	6.5	Storyboards	110
	6.6	Hands-On Activities	114
	References		123
7	**Design Fiction and Speculative Sketching**		125
	7.1	Sketching the Future!	125
	7.2	What Is Design Fiction and Speculative Design?	126
	7.3	Speculative Sketching for Ideation and Exploration	126
	7.4	Speculative Sketching for Visual Narratives	130
	7.5	Speculative Sketching for Artefacts and World Building	133
	7.6	Hands-On Activities	136
	References		140

8	**Accessibility of Sketches**		143
	8.1	Introduction	143
	8.2	Rationalisation	144
		8.2.1 Sketching in HCI Accessibility	144
	8.3	Sketch and Disability	145
	8.4	Additional Techniques for Accessible Sketches	146
		8.4.1 Alternative Text	146
		8.4.2 Alternative: AltNarrative	150
		8.4.3 Colour Contrast	153
		8.4.4 Typography	154
		8.4.5 Visual Style and Consistency	155
		8.4.6 Digitalising	155
	8.5	Hands-On Activities	156
	References		157
9	**Digital Sketching Techniques**		161
	9.1	Getting Started	161
	9.2	What's New, What's Different	162
	9.3	Miriam's Story	162
		9.3.1 Miriam's Practical Application Tips (for the Digitally Shy Sketcher)	166
	9.4	Makayla's Story	167
		9.4.1 Makayla's Practical Application Tips for Digital Sketching	168
	9.5	Hands-On Activities	173
	References		175
10	**Remote Sketching**		177
	10.1	Introduction	177
	10.2	Tools and Recommendations	178
	10.3	Remote Sketching Case Studies	180
	10.4	Hands-On Activities	189
	References		194
11	**Applying Sketching in Research and Practice**		195
	11.1	Destination, Sketching!	195
	11.2	Sketching and Networking	196
	11.3	Digital Sketching Footprint	198
	11.4	Sketching as a Research Method	201
	11.5	Publishing Your Sketches	205
	11.6	Visual Abstracts	210
	11.7	Hands-On Activities	213
	References		215
12	**Sketching with Other People**		217
	12.1	Those Who Sketch Together	217
	12.2	Why, Who, What, Where?	218

	12.3	Workshops and Events..................................	218
		12.3.1 Provide Materials!.............................	218
		12.3.2 Get People to Engage and Relax..................	220
		12.3.3 Lead by Example and Encourage...................	221
		12.3.4 Digital Workshops and Events....................	225
		12.3.5 Informal Sketching Events.......................	227
		12.3.6 Case Study: Example Sketching in HCI Event, Workshop, or Course Structure..................	228
	12.4	Graphic Recording and Visual Facilitation................	232
	12.5	Research with Other People............................	234
	12.6	Sketch Event Prompts.................................	239
	12.7	Sketch Ethics and Consent.............................	240
	12.8	Case Studies...	247
		12.8.1 Case Study #1: Icebreakers and Warm-Ups.........	247
		12.8.2 Case Study #2: Current Experience Comic Strips (CECS)............................	250
		12.8.3 Case Study #3: Conference-Level Sketching.......	254
		12.8.4 Case Study #4: Co-sketching in Research.........	255
	12.9	Hands-On Activities..................................	257
	References...		259
13	**The Future of Sketching**.................................		**263**
	13.1	Introduction...	263
	13.2	Teaching and Learning................................	264
	13.3	Possibilities for Sketching with Emerging Technologies.....	265
	13.4	Possibilities for Generative AI..........................	267
		13.4.1 Generative AI and Sketching.....................	268
		13.4.2 Sketching with Generative AI as a Collaborator....	269
	13.5	Sketch and Sustainability..............................	273
	13.6	Hands-On Activities..................................	274
	References...		277
14	**Additional Resources and Community**......................		**279**
	14.1	Welcome to the World of Sketching in HCI!..............	279
	14.2	Website and Social Media.............................	279
	14.3	Miriam's Favourite Personal Style......................	281
	14.4	Makayla's Personal Style..............................	286
	14.5	Observational Sketching...............................	293
	14.6	Photographing and Scanning Your Sketches...............	301
	14.7	Books and Media.....................................	304
		14.7.1 Miriam's Inspiration............................	304
		14.7.2 Makayla's Inspiration...........................	304
	14.8	The Sketching in HCI Manifesto........................	306
	14.9	Invitation..	307
	14.10	Hands-On Activities..................................	308
	References...		308

About the Authors

Makayla Miranda Lewis

By Day…

Dr Makayla Miranda Lewis (aka Maccy, only my mother calls me by my full name) is a Senior Lecturer in Computer Science (User Experience Design) at Kingston University London, United Kingdom (for a quirky New Scientist Jobs article about Makayla's workday, please visit https://www.newscientist.com/nsj/article/what-does-a-senior-lecturer-in-computer-science-do-/). Makayla has a PhD in Human-Computer Interaction from the City University of London and was inspired by her mother's physical disability, "Cerebral palsy, online social networks and change". Over the last 12 years, since completing their PhD, Makayla has researched and taught at various universities in England; her interests include human-computer interaction, user experience design, auto-ethnography, arts and creativity emerging technologies, and of course, SKETCHING with Dr Miriam Sturdee.

By Night…

Dr Maccy is also an accomplished visual thinker, sketcher and visual notetaker, and their works have been featured in three visual thinking books, an *Adobe* blog and several Human-Computer Interaction publications (and now this book). Maccy is an active scribbler, doodler and sketchnote who enjoys sharing their creative process at academic and industry sketching events courses and organising international meetups, e.g. *Sketchnote Hangout* and *Sketchnote LDN*. They are often found in meetings, seminars, meetups and on the sofa with Umbriel (a rather naughty but cute rabbit), Lyra (a little moody, but her button nose is so cute) and her guinea blobs (pigs) sketching her experiences, feelings and birds (they do not know why, but their default "I don't know what to draw" is birds). Maccy is a creative share; over 1,500 sketches can be littered across social media, websites and publications.

For more information about Dr Maccy, please visit: www.makaylalewis.co.uk.

Miriam Sturdee

Miriam is a lecturer and researcher in the School of Computer Science at the University of St Andrews in Scotland, UK. Following a varied career path, she initially studied psychology, worked in marketing, graphic design, publishing, illustration and undertook a Masters in Fine Art at Edinburgh College of Art. Later, bringing together all of her interests, she pursued an MRes and PhD in digital innovation at Lancaster University in 2013. It was during her PhD that she started investigating the role of sketching in designing for complex and novel technologies, especially shape-changing interfaces and tangible interaction. More recently, the sketching approach has transitioned into incorporating how arts-based approaches can inform technical fields within STEM (Science, Technology, Engineering, and Mathematics), with a particular focus on future-focused interactions. Her other research explores everything from humanised iPhones and subconscious user-experiences to TTRPG (Table-Top Roleplaying Games) in User Experience Design. As well as taking on regular service roles within the SIGCHI (Special Interest Group on Computer-Human Interaction) and other ACM (Association for Computing Machinery) communities, Miriam also teaches the long-running sketching in HCI (Human-Computer Interaction) course with Makayla Lewis, a journey that started in 2014 and has led to the publication of this book. Outside of work, Miriam herds cats, makes delicious food, sketches non-HCI things and loves being outdoors (in between rain showers).

Chapter 1
Introduction

1.1 Welcome!

Sketching is a universal activity but an often overlooked skill—yet it can benefit researchers and practitioners in Human-Computer Interaction (HCI)—sketching has proven to be a valuable addition to skill sets in academic and industrial contexts. Many individuals lack the confidence to take up sketching after years of non-practice, but it is possible to relearn, improve, and apply this skill in practical ways. This book is a sketching journey, from scribbles and playful interpretations, to helpful theory, storytelling, and practical applications. Individuals will learn techniques and applied methods for utilising sketching within the context of HCI.

Although celebrated in many forms, sketching in HCI is often relegated to a short section, or an addendum to another activity. This can be because the expertise to teach it is not available, or there are time or space constraints. By utilising this book however, we hope that educators and students alike can teach themselves practical sketching skills which cover a multitude of areas within HCI—and beyond—in computer science, interaction and user experience design, and other related fields.

This book is not what you expect—it does, of course, contain short essays covering different aspects of sketching, and practical activities to complete, but it is more than that. You will be joining us as we explore and practice the art of sketching in HCI through our own eyes and within our own environments (Fig. 1.1). Throughout the book you will find sketches of our surroundings, from our research activities, and examples from real-life courses we have held over the past 9 years.

No prerequisites are required, complete beginners are welcome, as are those who already sketch but are open to reimagining their practice and updating their skills within the context of HCI. *Anyone can learn to sketch*: it is a matter of starting, practicing, and developing your own personal style—you do not have to be artistic!

Most ideas are conveyed more effectively with sketch than words, and sketches are quick and inexpensive to create (and therefore also disposable when no longer

© The Author(s), under exclusive license to Springer Nature
Switzerland AG 2024
M. Lewis, M. Sturdee, *Sketching in Human-Computer Interaction*,
https://doi.org/10.1007/978-3-031-50136-4_1

Fig. 1.1 Welcome to our book! Makayla (left) and Miriam (right). *Procreate* App on *Apple iPad Pro* using *Apple Pencil*. Makayla Lewis, 2020

needed). Unlike detailed drawings and complex diagrams, sketches do not inhibit early exploration—there should be little investment in each one as they are timely and made in the moment. This also means that sketches are plentiful and you can entertain a huge number of ideas with multiple sketches for each one. By further annotating your sketches you can also make them accessible to users and stakeholders as part of a larger design process.

1.2 What *Is* a Sketch?

The traditional view of the "sketch" is that of a visual representation of an idea, or a short, fast drawing on paper—although it has more meanings depending on the context of use. In HCI, for example, the sketch can take on new roles as diverse as a section of code, a collection of actions, sounds, or even a rapid prototype. The pen and paper exemplar also has a new life within the context of computation, in that the sketch can be recognised as such using algorithms, converted from sketch to digital representation in 2D or 3D, and even used as an input method.

As much as sketching is universal in human culture, it also transcends research disciplines. Art lays the foremost claim, and via art, architecture and engineering emerge as primary users of sketching as a methodology and process. To "sketch" with a computer has been of interest since the early 1960s, with the earliest iterations using code to make lines (Sutherland, 1964) or proposing to involve direct "pencil" to paper interaction (Woo, 1964). The sophistication of current devices such as specialised digital drawing tablets and stylus bears testimony to these early ideas, with the ideal being coupling the intimacy and freedom of the sketched image with the power and possibility of the computer.

1.3 Why Use Sketching in HCI?

The act of sketching is a rite of passage in human development—we learn to use tools and make marks even before we learn to speak, and this kind of visualisation is a particularly human method of thinking, expression, and communication. So how can we reconcile this within the construct of Human-Computer Interaction?

If we look to historical examples, we can see the power of the sketch, for example, the *Alessi Juicy Salif* lemon squeezer started life on the back of a napkin (Alessi, 2016)! Famous physicist Richard Feynman was a prolific visual thinker and one of the greatest minds of our time. Sketching transcends the boundaries of discipline. Famous inventor Thomas Edison and his staff kept sketchbooks, over 2500 of them, each with several hundred pages…

Visualising real-world research can be daunting! An easy way to start is with a small part, and utilising the skills you will learn in this book can help. The primary ways in which you might describe the use of sketching in HCI are for documentation, reflection, and communication, although within HCI there are many ways a sketch can go further… (Fig. 1.2).

1.4 The Basics

1.4.1 Ideation

Start with thinking through sketching! Ideation is one of the universal constants in sketching in relation to research (we can also call it *brainstorming*) and forms part of almost every (if not all) discipline. Ideation in this context relates to the quick

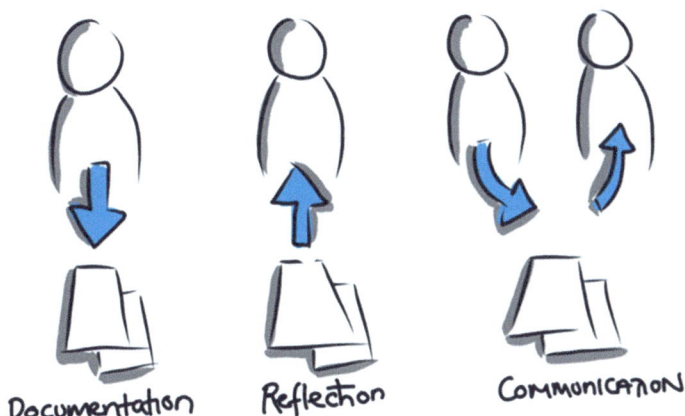

Fig. 1.2 *Sketchnote LDN* meetup mantra for visual thinking. *Photoshop* using *Wacom Cintiq Companion* and *Wacom* Pen. Makayla Lewis, 2018

generation of many different visual ideas and is utilised at the beginning of a project phase, and it is also used as part of a user-centred design process. When we start generating ideas, there may be the temptation to fixate on your first "reasonable" idea, but this could be a mistake! Try instead fast sketching many ideas, and don't be afraid to be silly—often the most crazy sounding ideas are the good ones—where you think outside of the box (or, as Clint Runge, from the *Interaction Design Foundation*, stated: "I try not to think out of the box anymore, but on its edge, its corner, its flap, and under its bar code…") (Fig. 1.3).

1.4.2 Iteration and Reflection

Sketching is perfectly suited to the iterative design process in HCI, where a prototype or image has been decided upon but requires refinement or further ideation. Iterative sketches in HCI relate to the incremental development of visual ideas, much as they would in design disciplines, but are used in relation to the prototyping process, on paper or on screen. Similar to ideation, the iterative sketching process is a visual record of ideas, but also reflection upon and the development of those ideas. In HCI research iteration is used to examine ways of exploring and analysing previously generated ideas (Fig. 1.4).

1.4.3 Evidence and Documentation

Sketched visuals, including iterations can be a form of record for processes and findings in a cheap, fast, and accessible manner. This kind of usage relates to the published findings and documentation relating to research projects in HCI and can be purely process related or show the results from computational output. Pictorials (Blevis et al., 2015) containing sketches are becoming more popular for documenting processes within HCI, and at conferences or other live events, sketchnoting and scribing are being embraced as adding value to both attendance and the legacy of events. Sketches can also be used as a documentary of meetings and collaboration, both in research and industry (Fig. 1.5) and in public as a form of expression on social media (Figs. 1.6 and 1.7).

1.4.4 Communication and Dialogue

Given the universal nature of the visual sketch, it is therefore a logical extension of the concept that it be used as a form of dialogue. Sketching can enhance the collaborative experience and can also be used as a way of communicating concepts remotely via digital pen/paper capture, where textual language cannot adequately

1.4 The Basics

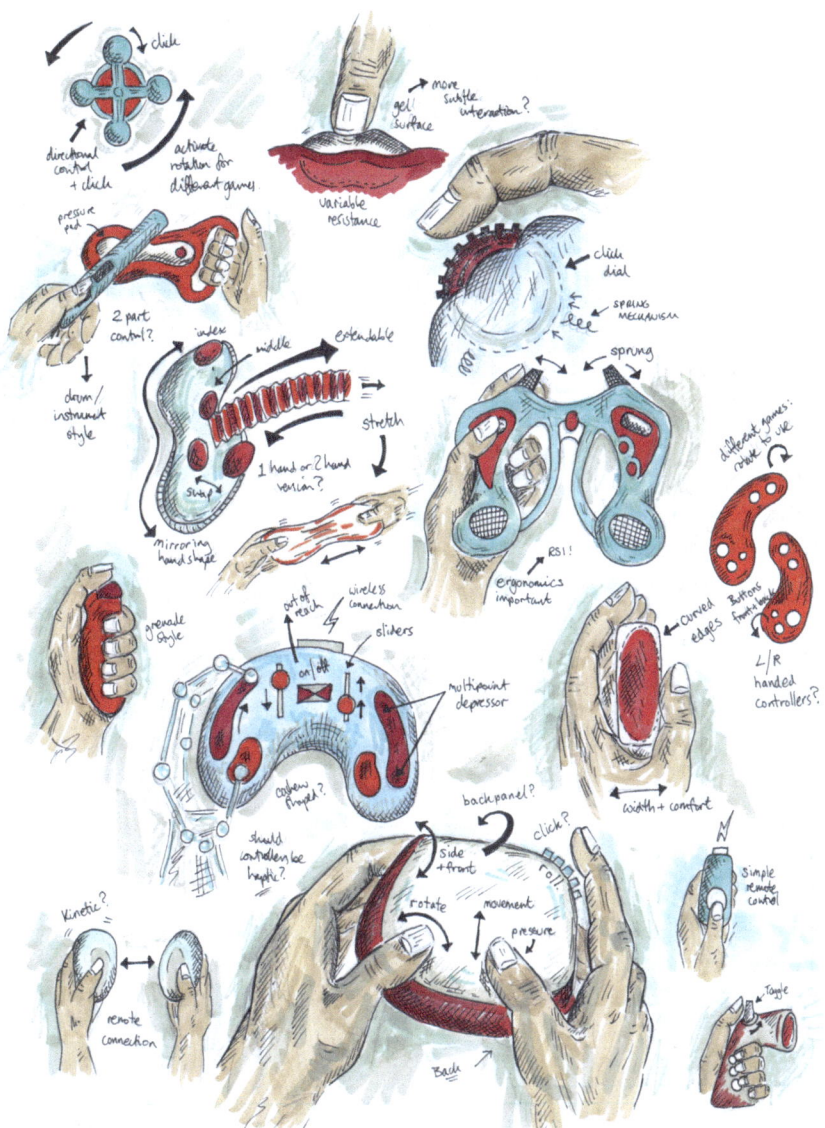

Fig. 1.3 Ideas for a game controller, including annotation to show interaction. Pencil, pen, fineliner, and Bristol board on paper. Miriam, Sturdee, 2016

demonstrate meaning—although, as we already said—annotations can help where explanation by visual means alone is insufficient. A sketch can also be a means to *open* dialogue, through a simple exchange. Some successful working relationships for us started with sharing a sketch on social media, and the ensuing conversations led to future collaborations! Don't be afraid to share sketches, regardless of how "neat and tidy" you think they are (Figs. 1.8 and 1.9).

Fig. 1.4 Iteration and elaboration on an idea for an application. Pencil on paper. Miriam Sturdee, 2016

1.4 The Basics

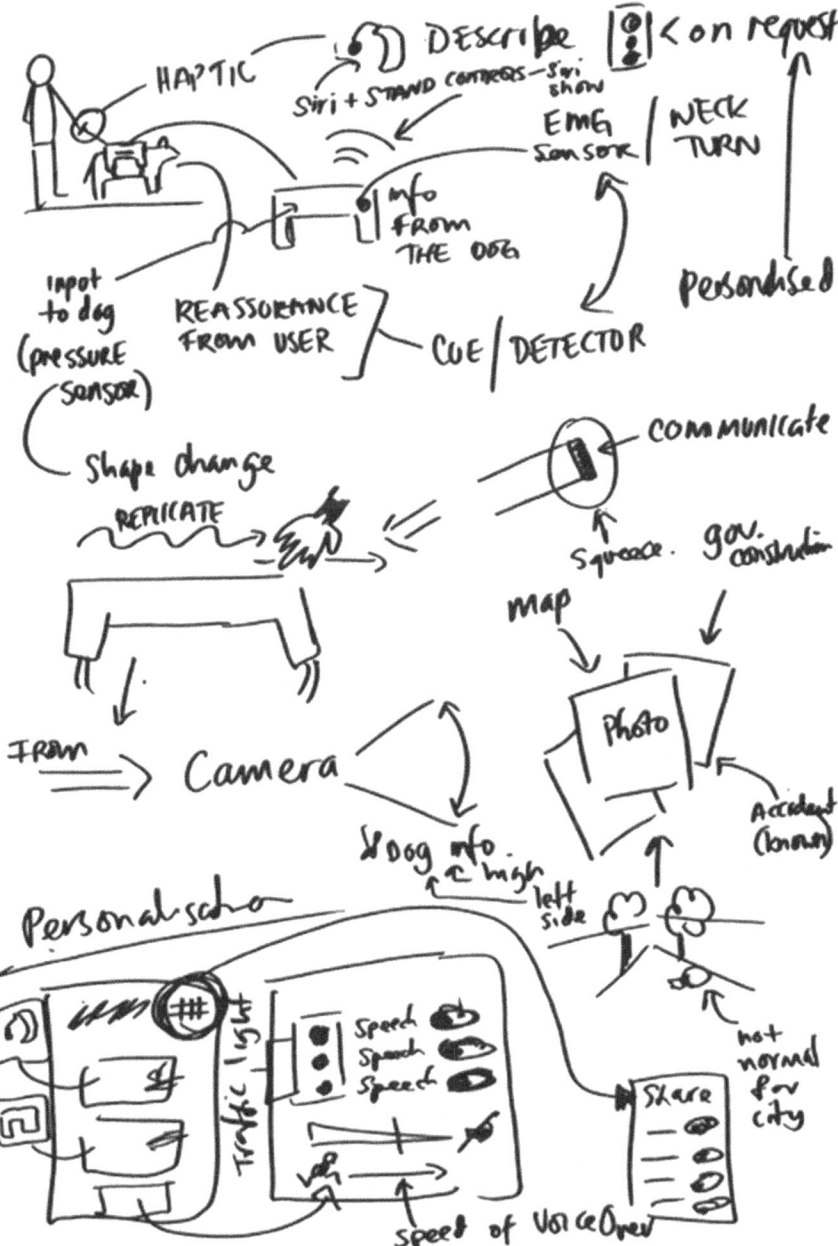

Fig. 1.5 Extract of lo-fi sketches of wayfinding app and guide dog sensor at a "designing for blind and visually impaired" workshop. *Photoshop* on *Microsoft Surface Pro* using *Microsoft Surface Pen*. Makayla Lewis, 2019

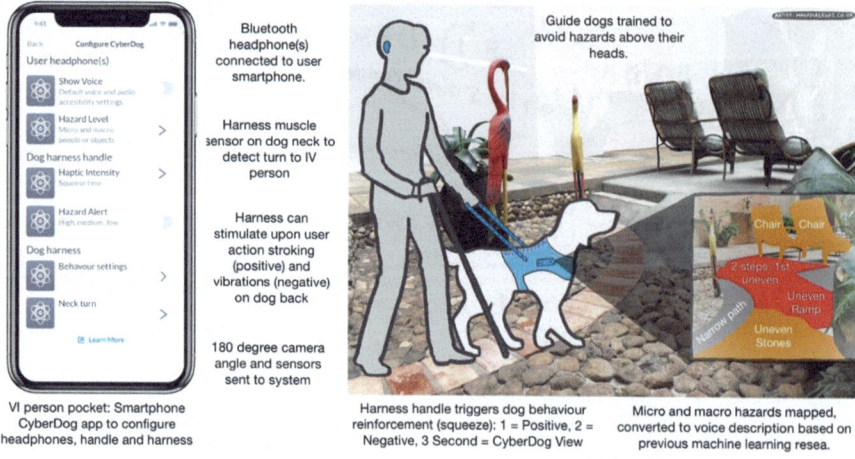

Fig. 1.6 Extract of mid-fi sketch of wayfinding app and guide dog sensor concept at a "designing for blind and visually impaired" workshop. *Photoshop* on *Microsoft Surface Pro* using *Microsoft Surface Pen*. Makayla Lewis, 2019

Fig. 1.7 Sketchnote from Japan Centre "Tokyo 1964 Designing Tomorrow". Fineliner pen and marker on paper. Makayla Lewis, 2021

1.5 The Complex

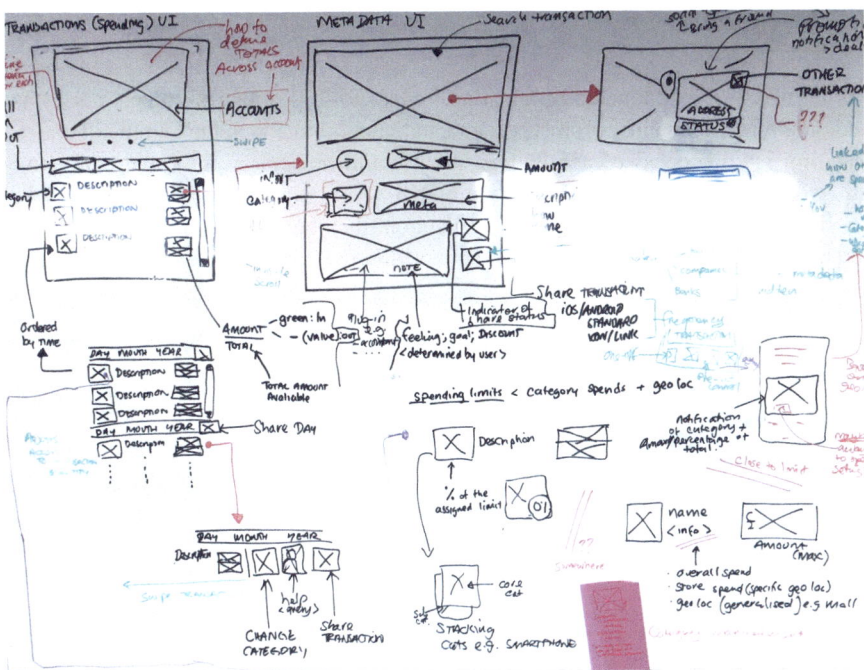

Fig. 1.8 Collaborative ideation for FinTech app. Dry marker on whiteboard. Makayla Lewis, 2019

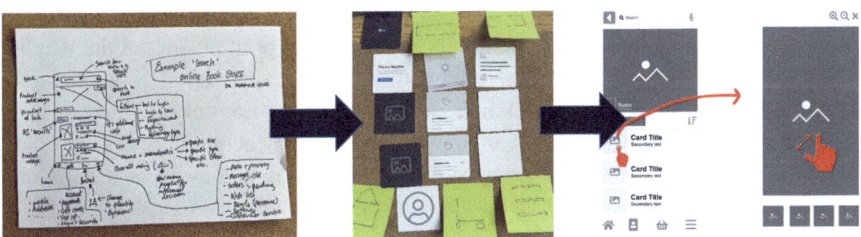

Fig. 1.9 Live demonstration of the role of sketching in the low-fidelity to mid-fidelity process as part of a Kingston University, London, MSc User Experience module – Design Thinking Theory and Practice. Fineliner pen on paper; fineliner pen on post-it notes; and *Miro* online whiteboard user experience wireframe elements. Makayla Lewis, 2023

1.5 The Complex

This book will teach you the basics of sketching, but if you are interested, sketching is being developed and used within HCI in many exciting ways. Some of these are described below, and we go into more detail in Chap. 11 where we look at the future of sketching in HCI.

1.5.1 *Elaboration*

Sketching can be conducted on existing items (not only other sketches, but as an addendum to documents, prototypes, and statements) to add value and to aid understanding via a process of annotation. What could be seen as a simple annotative process can take on new meaning within HCI—more than simply making notes or doodles on a text—when such sketched annotations become interactive and transferable, or can be transmitted across the world in real time.

1.5.2 *Input*

Direct input is a use of sketching that is novel to HCI and computing, in contrast to traditional, freehand sketching where the resulting image is a visual output only. The sketch-as-input relies upon complex computational processes to recognise lines, shapes, distances, and stroke widths—as well as intended meaning. The sketch-as-input is an essential part of sketch-based interfaces and is made possible by software allowing for sketch recognition.

1.5.3 *Output*

The visual output of freehand sketching is an easy concept to grasp, but the way in which sketching output can be generated within HCI is enhanced in comparison to art and design processes. The sketch as output can relate not only to an image on paper or screen, but how sketches can now be generated by neural networks or other programs utilising a variety of imagery and rendered as if they were a freehand representation for both generic scenes and portraits of people.

1.5.4 *Tool*

Tool-based sketching is an intermediary concept, where sketching is utilised as a means to an end. This relates to the sketch as having a purpose beyond simple input or output—rather, it provides a service in a specific context, such as being a tool for increasing interactivity in paper prototyping or to direct movement.

1.6 A Short History of Sketching in HCI

The first mention of sketching in relation to computing can be found in a search of articles from 1964 in which we find Conn and Von Holdt (1964) discussing curve sketching and Ivan Sutherland's papers on the *Sketchpad* system (Sutherland, 1964). Interestingly, at first, the hand-drawn sketch was seen as the gold standard with Keller (Keller, 1965) postulating the value of using hand sketching of Burmester curves to estimate studies before a computer was used and to check the data was accurate afterwards. Early research combining the sketch and computers was focused largely on mathematical calculations.

In 1975, Taggart proposed sketching as an "informal dialogue between designer and computer" (Taggart, 1975), and in terms of the technical elements of sketching-computer dialogue, Negroponte (Negroponte, 1973) pioneered sketch recognition and search in the early 1970s. Then, with the launch of SIGGRAPH in 1974, the computer as an essential part of graphics production was born, and by 1976, researchers started looking into computer-assisted sketching—using the WHATISFACE SYSTEM (Gillenson & Chandrasekaran, 1975). The first Special Interest Group on Computer-Human Interaction (SIGCHI) conference in 1982 marked a seminal moment for HCI, though there was little written on sketching until Preece's work using an interactive graph sketching program for education (Preece, 1984) and Miller's exploration of sketching for page layouts using computers in 1984 (Miller, 1984)—part of the beginning of desktop publishing.

By 1990 the exploration of sketching had gained in popularity, alongside computer-aided design (CAD) programs, which were intended to replace hand-drawn schematics for engineering and industrial design, although the two disciplines have frequently been examined together, such as comparing technical sketching and computer-aided design (Ryan, 1990), or more recently, with the suggestion that hand-drawn sketching has advantages over CAD (Veisz et al., 2012). One of the most vocal proponents of hand-drawn sketching is Gabriela Goldschmidt; although her works fall firmly within the design domain, the insights are applicable to HCI in that they provide justification and explanation of the processes, skills, and value of sketching by hand (Goldschmidt, 1991; Goldschmidt, 2017). Later years have seen work on sketch retrieval and collaborative sketching, 2D-3D conversion of sketched images, sketch-based passwords, and even robots that sketch. Sketching in HCI is in its Renaissance, and of course, now is the perfect time to start learning!

1.7 How to Use This Book

Each chapter will have an introduction to the area covered, with sketches from both authors (Makayla and Miriam). Some sketches are created especially for this book, to provide detailed examples of the topics at hand, and others are taken from our in-person or virtual courses, or from our personal archives. These sketches provide examples of the background theme but also illustrate how a completed activity might look (although they are guidelines only—we expect students to develop their own style during the course of this book).

In each chapter we will introduce different approaches for developing your practice within the outlined area, with examples from real-world contexts, followed by practical, hands-on activities to guide you through the process from start to finish. You can do each hands-on activity as you read or complete the activities as stand-alone exercises. At the end of each chapter's introduction, we also provide tips for extending practice and exploring each topic in your own time.

You can also visit our website which accompanies this textbook (W1) and find more activities, news, and research in this area.

And REMEMBER... (Fig. 1.10)!

1.8 Materials

For each chapter, we suggest having a choice of pens and paper. For the earlier chapters, any scrap of paper is fine—the back of an envelope, or a piece of printer paper. You can use any pen, biro, or pencil you have to hand, although we recommend a black fineliner with an easy flow. We do suggest you use a pen, not pencil, as it prevents abandonment of ideas, and there is also an increased chance that you'll finish the sketch (sense of commitment). As you develop your practice, you may also wish to get a sketchbook and record the progress of the course (see the next section).

We recommend the following basic materials for this book (e.g. Figs. 1.11 and 1.12):

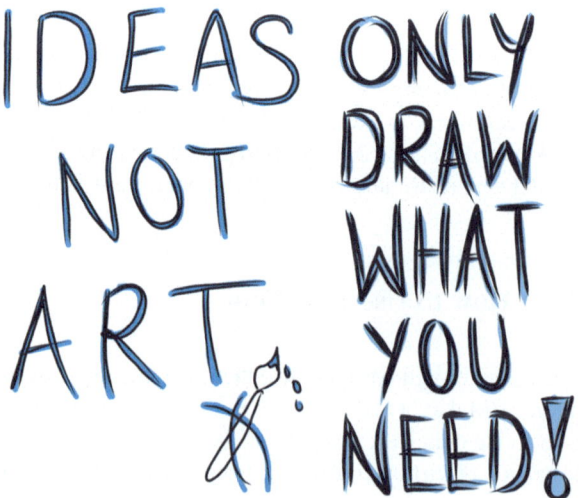

Fig. 1.10 The *Sketchnote LDN* and Sketching HCI mantra. *Photoshop* on *Wacom Cintiq Companion* using *Wacom* Pen. Makayla Lewis, 2016

1.8 Materials

Fig. 1.11 Exemplar of traditional sketch materials: plain paper sketchbook, pencil, fineliner pen (black), eraser, grey marker, and a light-coloured marker. *Procreate* App on *Apple iPad Pro* using *Apple Pencil*. Makayla Lewis

- Plain white paper: A5 printer paper or A5 size or smaller sketchbook, e.g. hardback non-spiral, is recommended as it will provide a hard surface to sketch on when not at a table or desk.
- Several packs of Post-it notes, colours are your choice.
- A series of black drawing pens with easy flow ink, e.g. UniPin 0.1, 0.5, 0.5, and 0.7.
- Mechanical pencil, e.g. 0.7.
- Two pastel-coloured markers of your choice to highlight important areas, e.g. *Copic* (alcohol-based markers) or Tombow ABT brush pen (water-based marker).
- Grey marker to add depth, e.g. *Copic* C3, C4 or Tombow ABT N75.

You may also wish to use a digital drawing device/tablet if preferred (e.g. Fig. 1.13):

- Digital drawing tablet, e.g. *Apple iPad Mini/Air/Pro*, *Samsung Galaxy Tab S*, or *Microsoft Surface Go*.
- Digital drawing pen, e.g. *Apple Pencil*, *Samsung S Pen*, or *Microsoft Surface Pen*.
- Digital drawing app, e.g. *Procreate* App on *iOS* and *Autodesk Sketchbook* on *Android* and *Windows 11*.

Fig. 1.12 Sketch of a *Copic* marker posted to Twitter (now *X*). Fineliner and marker on paper. Miriam Sturdee, 2019

If you are running a class and using this book for reference, we suggest acquiring the following items (~35 students):

- 500 sheets A4 white printer paper (printer pack).
- 3 sets of 12 black fineliner pens.
- 6 blocks of sticky notes in a variety of colours (~90 notes per pad).

At the end of the book, we will discuss and list additional resources, including art materials we recommend for extending and refining your practice, extra activities, and further reading.

1.9 Keeping a Sketchbook

Keeping a sketchbook (sometimes referred to as "sketchbooking") can have many purposes:

- Capture and explore ideas (to create), e.g. Fig. 1.14.
- Gather inspirational materials, examples, and observational photographs. Don't forget to note the source (attribute) inspirations not generated by yourself.

1.9 Keeping a Sketchbook

Fig. 1.13 Exemplar of digital sketch materials: *Procreate* App on *Apple iPad Pro* using *Apple Pencil*. Makayla Lewis

- Develop sketching skills, e.g. using the activities presented in this book, e.g. Figs. 1.15 and 1.16.
- Document ideas and thoughts, e.g. Figs. 1.16, 1.17, and 1.18 (also see Peterman & Peterman, 2015).
- Act as a memory aid for your thoughts, feelings, and knowledge exchange events/lecturers/seminars, e.g. Fig. 1.7 and Lamm (2015).
- Explore and test different sketching media and styles, e.g. Fig. 1.13.
- Creative dump (visual journal) of your everyday thoughts and growth over time (a record of life), e.g. see Sokol (2019) and Quitely (2019).

A sketchbook is not one, some, or all of these purposes. Stay calm by the opportunities for use present here; as the sketcher, you should let the purpose develop as you work through this book. There is no right or wrong way to use a sketchbook; it

Fig. 1.14 Exemplar of traditional sketchbook page using mixed media. Colour pencil, marker, watercolour, gouache paint, and fineliner on paper. Makayla Lewis, 2023

is a versatile tool to enhance your creative expression. There are no mistakes but opportunities; as Bob Ross once said:

"There are no mistakes, just happy accidents"
and
"Talent is a pursued interest. Anything you're willing to practice, you can do" (Calm Blog, 2023, W2).

1.10 Practical Application Tips

During the introduction to each chapter, we may also include "practical application tips". These tips could be theory (*only draw what you need!*), specific to the item you are sketching, related to the medium you are using, or the situation in which you find yourself sketching. They will likely also relate directly to the hands-on activities at the end of each chapter. So for this section:

Practical Application Tips
- Be kind to yourself—you will get more confident over time.
- Enjoy yourself—sketching should be fun. If you are getting frustrated, step away from the page and come back to it later.

Fig. 1.15 Exemplar of traditional sketchbook page using. Marker and fineliner pen on paper. Miriam Sturdee, 2022

1.11 Hands-On Activities

At the end of each chapter (or section—in larger chapters), there will be around three or four practical, hands-on activities to complete. You may dive straight into these and read the chapter introduction later or simply complete each on its own. Either way is fine.

For some activities we will also follow the instructions and sketch alongside you.

Activities are for either individuals or pairs/groups and will be broken down as per this example:

Activity #: TITLE (Individual/Pair/Group Activity)
Learning objective—What the student should achieve after completing this activity.

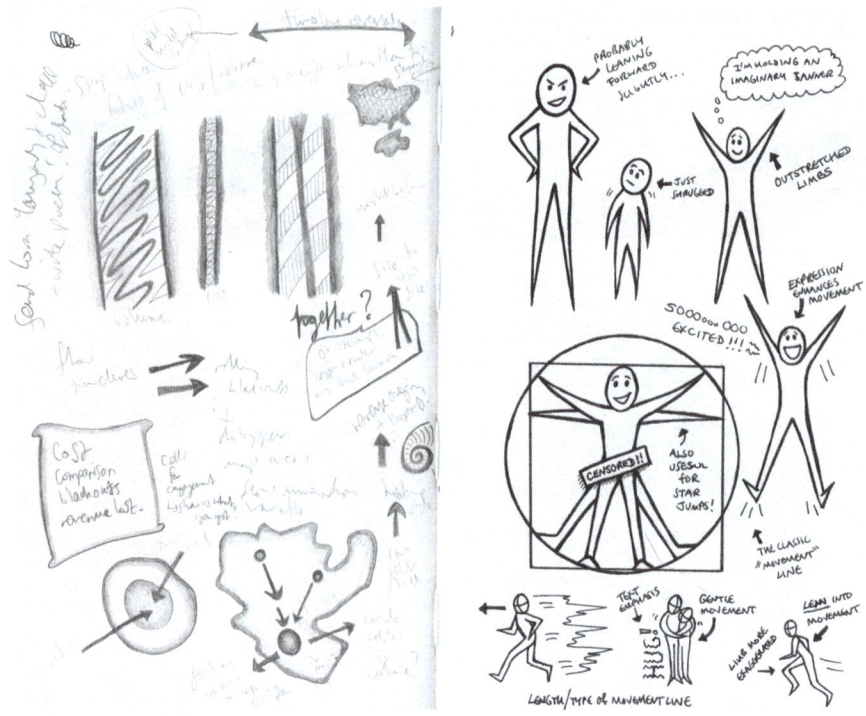

Fig. 1.16 Sketchbook: contrasting events and explorations on facing. Pencil and fineliner marker on paper. Miriam Sturdee, 2018

Fig. 1.17 Exemplar of digital sketch materials "Bird". Kindle Scribe and Kindle Scribe Pro. Makayla Lewis, 2023

Time—Activities are designed to be short and achievable within the scope of a workshop or class during which you may have other information to cover. The longest activities will be around 30 minutes and usually involve group work. You may also complete the activities as homework.

Materials—Basic materials needed and any extra items that may be required (e.g. sticky notes, rulers).

Procedure:

- A bullet pointed list clearly explaining each step.
- This will be followed by an example sketch for reference or linked to an example in the main text.

After the procedure for each activity, we may add our own examples and descriptions, or extra notes!

1.12 References and Resources

Each chapter will have references that appear in the text, and these are broken down into the following: (1) *Books and Papers*, either academic resources or other texts referenced in the introduction; (2) *websites*, digital sources for students to read about items or themes mentioned in the text and see image examples; and (3) *further reading*, for those who wish to delve into the chapter theme in more detail. Also, a quick note on the captions for our images: Where images are from our personal archives, we state the year the image was created, alongside our names; where the image was created by us, but has been included in a published work, we state the reference; if an image has been created especially for this book (as many are!) we simply put our name at the end with no date.

1.13 A Fun Hands-On Activity for Reflection

To map your progress, we would like you to sketch out, on a single page, your mental model of what HCI looks like. We know you are just starting, but this will enable you to compare your progress across the course of the book. Don't worry—any scrawl, doodle, or stick figure will do! We'll ask you to do this again in Chap. 11. Remember to keep your sketch safe somewhere, and don't look at it until you have finished all of the chapters!

Activity 1.1: What Does HCI Look Like? (Individual Activity)
Learning objective—To understand and document your progress and achievements in sketching in HCI
Time—5–10 minutes
Materials—A4 paper, black fineliner, coloured pens or pencils if desired

Procedure:

- Draw your understanding or current perception of HCI as a field of study.
- Don't overthink this; just get something onto the page, quickly and without worrying about mistakes.
- We don't expect you to be an accomplished sketcher at this stage, and you do not have to show anyone else this sketch—if you are in class, you can put this away or give it to your lecturer or professor to keep somewhere until the end of the course.

Figure 1.18 is an example from Makayla, using a single-line scrawl technique. She has drawn an abstract self-portrait in response to pushing sketching into HCI. Dynamic lines that interconnect show a side view of Makayla holding a circular object, her sketches, and art, being pushed into another digital version of

Fig. 1.18 Trying to find a space for sketching and arts in HCI. *Procreate* App on *Apple iPad Pro* using *Apple Pencil*. Makayla Lewis, 2022

1.13 A Fun Hands-On Activity for Reflection

her; this person has human and robotic features. Frustrated links are circling all over the page.

In contrast, Fig. 1.19 is an example from Miriam, who enjoys layering colour and line to build up an image. This image shows how HCI is about people, not technology, and how we need to focus on that rather than computing as a solo field.

Fig. 1.19 What does HCI look like? *Procreate* App on *Apple iPad Pro* using *Apple Pencil*. Miriam Sturdee, 2023

Note that our styles are completely different! Also, please do not negatively compare your sketch to ours, we've been doing this for a long time...
Now you are ready to begin your Sketching in HCI journey!

References

Books, Papers, and Articles

Alessi, A. (2016). *The dream factory: Alessi since 1921*. Rizzoli.
Blevis, E., Hauser, S., & Odom, W. (2015). Sharing the hidden treasure in pictorials. *interactions, 22*(3), 32–43.
Conn, R. W., & Von Holdt, R. E. (1964). *Curve sketching by digital computer* (No. UCRL-12226). Lawrence Radiation Lab., Univ. of California.
Gillenson, M. L., & Chandrasekaran, B. (1975). A heuristic strategy for developing human facial images on a CRT. *Pattern Recognition, 7*(4), 187–196.
Goldschmidt, G. (1991). The dialectics of sketching. *Creativity Research Journal, 4*(2), 123–143.
Goldschmidt, G. (2017). Manual sketching: Why is it still relevant? In *The active image: Architecture and engineering in the age of modeling* (pp. 77–97).
Keller, R. E. (1965). Sketching rules for the curves of Burmester mechanism synthesis. *Journal of Manufacturing Science and Engineering, 87*, 155.
Lamm, E. (2015). *Sketchnotes 2013/2014* (1st ed.). CreateSpace Independent Publishing Platform.
Miller, M. (1984). *Spatial context as an aid to page layout: A system for planning and sketching* (Doctoral dissertation, Massachusetts Institute of Technology).
Negroponte, N. (1973, June). Recent advances in sketch recognition. In *Proceedings of the June 4-8, 1973, national computer conference and exposition* (p. 663).
Peterman, S., & Peterman, S. E. (2015). *The sketchbook project world tour*. Chronicle Books.
Preece, J. (1984). A study of pupils' graph concepts with a qualitative interactive graph sketching program. *Computers & Education, 8*(1), 159–163.
Quitely, F. (2019). *Drawings + sketches*. BHP Comics.
Ryan, D. L. (1990). *Technical sketching and computer illustration*. Prentice Hall.
Sokol, D. D. (2019). *A world of artist journal pages: 1000+ artworks| 230 artists| 30 countries*. Abrams.
Sutherland, I. E. (1964). Sketchpad a man-machine graphical communication system. *Simulation, 2*(5), R-3.
Taggart, J. (1975). Sketching, an informal dialogue between designer and computer. In *Computer aids to design and architecture*. Petrocelli/Charter.
Veisz, D., Namouz, E. Z., Joshi, S., & Summers, J. D. (2012). Computer-aided design versus sketching: An exploratory case study. *AI EDAM, 26*(3), 317–335.
Woo, P. W. (1964). A proposal for input of hand-drawn information to a digital system. *IEEE Transactions on Electronic Computers, 5*, 609–611.

Websites

W1 Our very own website! www.sketchingHCI.com
W2 Calm Blog, 2023. The 10 greatest Bob Ross quotes of all time. www.calm.com/blog/the-10-greatest-bob-ross-quotes-of-all-time

Further Reading

Bower, S. (2016). *The urban sketching handbook: Understanding perspective: Easy techniques for mastering perspective drawing on location.* Quarto Publishing Group USA.

Brand, W. (2019). *Visual doing: A practical guide to incorporate visual thinking into Your daily business and communication.* Laurence King Publishing.

Gaiman, N., & Riddell, C. (2018). *Art matters: Because your imagination can change the world.* Headline.

Greenberg, S., Carpendale, S., Marquardt, N., & Buxton, B. (2011). *Sketching user experiences: The workbook.* Elsevier.

Hoffmann, A. R. (2019). *Sketching as design thinking.* Routledge.

Roam, D. (2009). *Unfolding the Napkin: The hands-on method for solving complex problems with simple pictures.* Penguin.

Sturdee, M., Lewis, M., & Marquardt, N. (2018). Feeling SketCHI? The lasting appeal of the drawn image in HCI. *Interactions, 25*(6), 64–69.

Chapter 2
The Humble Line

2.1 Introduction

In geometry, a line is defined as a one-dimensional with undefined thickness (weight), or length (Fig. 2.1). A line can overlap or run parallel to itself, and with a dash of experimentation, it can form shapes that are sometimes abstract—although most of the time, they represent something the sketcher has experienced, felt, or wished for.

For those wishing to sketch, it starts with a blank page. For those who are not regular doodlers, this can be daunting—but it doesn't have to be. Forget still-life sessions at school or being asked to draw your own hand. Here, anything goes:

- Embrace your inner child.
- Start with a simple line, take it for a walk, and close your eyes if you need to.
- There. The page is no longer blank and you have begun (Fig. 2.2).

We were all born as artists, scribbling on the walls, sketching in the sand or dirt, or turning eating dinner into an abstract artwork, but somewhere along the way, we are told that we are not good enough, not conforming to the notion of what "art" is. However, that is not the reason or meaning behind a sketch. A sketch can be a piece of art, but more than that, a sketch starts a conversation, and a sketch can be the first block in building a visual world.

By the end of this chapter, you should be able to:

1. Get warmed up and get used to the feel of pen on paper, there are no mistakes!
2. Get used to drawing simple shapes, lines, and cross-hatching.
3. Appreciate and reinterpret abstract lines and doodles.

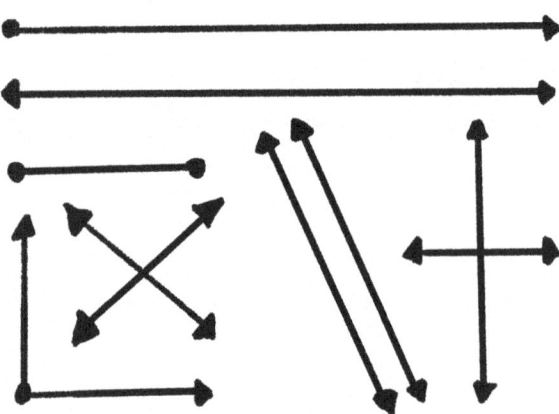

Fig. 2.1 Geometry of the humble line: point, extending a point, collection of points that extend, segment with two points, intersecting lines, angle where two lines share a point, parallel lines, and perpendicular lines. App *Apple 1 Procreate* App on *Apple iPad Pro* using *Apple Pencil*. Makayla Lewis, 2022

Fig. 2.2 Embracing our inner child by embracing the humble line for ACM CHI 2022 Conference. *Miro* online whiteboard on *Apple iPad Pro* using *Apple Pencil*. Makayla Lewis, 2022

2.2 Forget Everything You Learned at School

There is research that suggests that the ability to sketch and draw forms along similar pathways in the brain as does language (Cohn, 2012), and as natural linguists, it stands to reason that we are therefore natural artists. If you needed a justification to get back into mark making and visual expression, this is it—you never lost the ability, you never even needed to learn it. The problem is the expectations of others and social pressure. There are artists who sketch in a playful, loose way such as Quentin Blake (W1), or those who paint with the joy of simplicity such as in Naïve Art (W2). Photorealism (W3) (paintings or drawing that are indistinguishable from a photograph) is a skill possessed by a small minority, and whilst extremely skilled and precise, the technique does not necessarily showcase the personality and interests of the individual, e.g. Fig. 2.3.

As you become more in tune with your pen and the lines it can create, the more detailed your visualisations will become. The leap from doodle to majestic visual representation, although it will take practice, will challenge what it means to recreate one's experiences, feelings, and wishes on paper or digital canvas, e.g. Fig. 2.4.

To see the power of the *humble line*, view your surroundings. Even three-dimensional objects—if you look at them—are a series of lines. Ask yourself:

- What are the predominant lines that make up the object?
- What feeling does the object convey?
- What is the story behind the object?

For example, Makayla's modded Nintendo Gameboy Colour™ is made up of seven lines (Fig. 2.5) and offers an opportunity to reminisce about her 1990s childhood, the many hours of wandering amongst endless environments and capturing mythical creatures.

Practical Application Tips

- Keep a pen or pencil and paper to hand as often as possible.
- Practice by doodling during online meetings (doodling promotes active listening).
- Look over previous doodles and mark making to see if there are similarities or differences—have you changed your style or developed recurrent imagery?
- Experiment with different pens, pencils, and markers when warming up.
- If you have a tablet, try warming up digitally—consider how it differs.

Now try warming up and mark making yourself by completing Hands-On Activity 2.1.

Fig. 2.3 Humble line portrait. *Procreate* App on *Apple iPad Pro* using *Apple Pencil*. Makayla Lewis, 2021

2.3 Hands-On Activities

Activity 2.1: Warming Up (Individual Activity)
Learning objective—To make a start with sketching and gain in confidence with basic mark making
Time—5 minutes
Materials—A blank piece of paper or page of a sketchbook, any pen or pencil available
Procedure:

- Put pen to paper and draw a wavy line ("taking a line for a walk").
- Scribble without thinking about it; get to know the feel of the pen.
- Close your eyes and continue to scribble.
- Sketch some short straight lines.

2.3 Hands-On Activities

Fig. 2.4 The End of the Line. Marker and fineliner on paper. Miriam Sturdee, 2022

- Sketch some wavy lines.
- Fill in all the available space.
- Doodle in the larger gaps.

Our examples are below—remember there is no correct way to do this—warming up can be completely freestyle; just let go! (Figs. 2.6 and 2.7)

Activity 2.2: Shapes and Lines (Individual Activity)
Learning objective—Refine your lines; add simple shapes and textures
Time—5 minutes
Materials—A blank piece of paper or page of a sketchbook, any pen or pencil available
Procedure:

- Sketch some simple shapes—squares, rough circles (don't focus on perfection—nobody draws perfect circles), add some spirals and zigzags.
- Cross over some of your lines to create "cross-hatching" (basic grids).

Fig. 2.5 Finding lines in everyday objects. *Procreate* App on *Apple iPad Pro* using *Apple Pencil*. Makayla Lewis, 2022

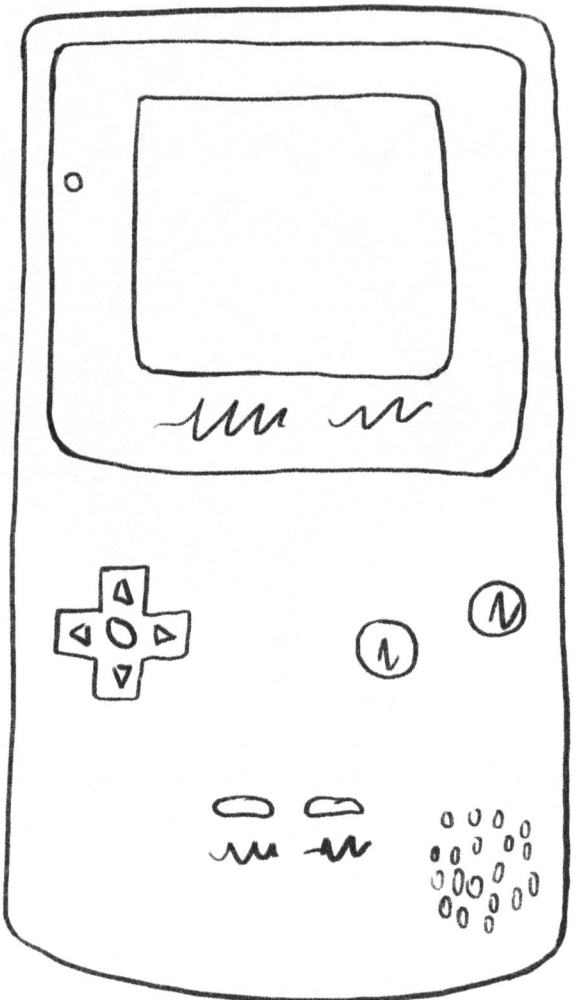

- Add textures to your page, e.g. make some dots in a group, overlay them, and add some which are more spread out; you can also keep drawing squares or circles or lines and keep building them up.
- Without looking at your hand or the paper, sketch four squares in a row, then four circles.
- Sketch a stick-person face and add a body.
- Fill in the rest of the page, any way you wish, using shapes and lines that feel natural to you.

Here, Makayla has used different colours to outline the stages we suggested for you to follow so you can see how the image has been built up (Fig. 2.8). Miriam's approach is more regimented, giving sections to each part (Fig. 2.9). How did

Fig. 2.6 Example of warming up. *Procreate* App on *Apple iPad Pro* using *Apple Pencil*. Makayla Lewis

you approach filling your page? This can tell you a lot about your preferred way of working with imagery.

Activity 2.3: Pareidolia Doodling (Group Activity)
Learning objective—Creatively interpret lines and turn them into an object, scene, figure, or animal
Time—10 minutes
Materials—A blank piece of paper or page of a sketchbook, any pen or pencil available
Procedure:

- Form pairs or small groups of 3–4 people.
- Place the tip of your pen somewhere on the piece of blank paper.
- Close your eyes and sketch a scribble that moves around and even crosses over itself.
- Open your eyes, pass your paper to your partner or member of your group; you should receive a scribble from someone else.
- Without over thinking, turn the scribble into something concrete, e.g. a dog.
- Pass the paper back to the person who drew the scribble.
- Repeat the exercise for each person in your group, or if working in pairs, create four doodles each.

Fig. 2.7 Example of warming up. *Procreate* App on *Apple iPad Pro* using *Apple Pencil*. Miriam Sturdee

2.3 Hands-On Activities

Fig. 2.8 Shapes and lines demonstrated in different colours: Part 1 black, part 2 red, part 3 blue, part 4 blue. *Procreate* App on *Apple iPad Pro* and *Apple Pencil*. Makayla Lewis

Fig. 2.9 Example of shapes and lines (single colour). *Procreate* App on *Apple iPad Pro* using *Apple Pencil*. Miriam Sturdee

We have done this exercise in two colours so you can see the original lines and the new lines that create the image. Often, these pareidolia doodles are figurative, but some people see landscapes and objects or simply create mini-artworks. Makayla has drawn a person in a tall hat and linked it up with the doodle on the left, which represents an artwork hanging on a wall that the person is looking at (Fig. 2.10). Miriam has drawn two people holding a vial of combustible science liquid, and an unrelated image of a dragon (Fig. 2.11). This is also a great icebreaker activity if you are ever running a workshop or event!

Fig. 2.10 Example of pareidolia doodling. Miriam Sturdee (black) and Makayla Lewis (blue). *Apple iPad Pro* using *Apple Pencil*

Fig. 2.11 Example of pareidolia doodling. Makayla Lewis (black) and Miriam Sturdee (green). *Apple iPad Pro* using *Apple Pencil*

Now that you have warmed up, you are ready to move on to the next chapters, where we will look at building your visual practice, with objects, people, actions, and more! Remember, you can warm up for every chapter or complete any exercise to help get in the mood for sketching.

References

Books, Papers, and Articles

Cohn, N. (2012). Explaining 'I can't draw': Parallels between the structure and development of language and drawing. *Human Development, 55*(4), 167–192.

Websites

W1 Personal website of the artist Quentin Blake – www.quentinblake.com/
W2 Tate Modern Gallery, London, Naive Art – www.tate.org.uk/art/art-terms/n/naive-art
W3 Tate Modern Gallery, London, Photorealism – www.tate.org.uk/art/art-terms/p/photorealism

Further Reading

Brown, S. (2015). *The doodle revolution: Unlock the power to think differently*. Penguin.

Chapter 3
Seeing the World in Icons

3.1 Introduction

Why are icons important? Part of the joy of icons is their simplicity, and their ability to convey meaning with great economy of line. Not only this, but well thought out, simple icons can provide a means of communication without words. This chapter will guide you through the process of developing your visual practice and creating a library of go-to icons for use in your communications materials, storyboards, and interface designs, and so forth.

Before written language, people communicated with imagery—for example, the Egyptian hieroglyphics that we find on the walls of tombs and monuments. Some modern alphabets even have icons as their roots, such as Chinese and Japanese Kanji, for example, the Kanji for "river" in Japanese shows wavy lines like water flowing, or the Kanji for "mountain" shows a series of peaks. These Kanji can also offer an insight into developing icons and images in our own practice—the simplification of a concept is key.

Proponents of visual language also include Otto and Mary Neurath, whose work on visual education and exploration is still relevant today (Neurath et al., 2010). They developed Isotype (International System of Typographic Picture Education), with the idea that complex ideas and images could be simplified into sketches, which could then be standardised. There are also multipurpose online icon libraries you can visit and use (e.g. *The Noun Project*, W1), but the most meaningful and useful icons and visuals will be the ones you make for yourself, reflecting your ideas, influences, and potential communications and outputs.

By the end of this chapter, you should be able to:

1. Identify relevant concepts and items for visual representation based upon your own studies and practices.
2. Sketch and solidify these concepts and items into simple icons or images, identifying where and when to simplify or use detail.

Fig. 3.1 Collaborative icon library from CHI 2019 conference course "Introduction to Sketching in HCI", fineliner pen on Post-it notes, Miriam Sturdee and Makayla Lewis, 2020

3. Refine and develop these concepts and items with practice and build your own, personally relevant, visual library.

Below we introduce different approaches for developing your practice in this area, with examples from different contexts, followed by practical, hands-on activities to guide you through the icon process from start to finish (Fig. 3.1). Remember, you can do each hands-on activity as you read or complete the activities as stand-alone practicals later!

3.2 Basic Visual Icons

Every day, explicitly or implicitly, we experience visual icons, e.g. a toilet symbol painted on a door, the pound sign above an automated teller machine (ATM), and the play/pause on a remote control. Visual icons are simplified symbols that represent objects that are embedded in every facet of our day-to-day lives. They are simple visual representations of our world that are created to be universally and consistently understood regardless of the language one speaks. Think of simple cave paintings, created hundreds of thousands of years ago by people who do not speak the same language as us today yet when we view them we understand, on a

Fig. 3.2 Basic icons. *Procreate* App on *Apple iPad Pro* using *Apple Pencil*. Makayla Lewis, 2021

rudimentary level, the objects they hold dear, e.g. knives made from rocks and bone to aid hunting and cooking.

You may be reading this book at home, in a library, or coffee shop. Look around and investigate your space; consider the key objects that represent the space and your person; these could be the keys to your home, an open book you are currently reading, or the clock on the wall that is counting down your working day. What are the key lines that represent that object? What lines or information can you reduce or remove without losing the essence of the object? Essentially, how can you represent that object simply? Here are a few examples (Fig. 3.2).

Practical Application Tips

- Practice by drawing from real objects whenever possible.
- Truly observing what you are sketching, often referred to as observational drawing, knowing what to keep and remove whilst maintaining the object essence can only be achieved if you know what the object in the real world looks like.
- Consider the perspective of the object; do not worry if it's not perfect, for example, consider the opening of a cup is oval not circular; be aware of ellipses.
- Use mark making to convey the object's surface and texture, lots of dots to represent concrete, hashes for bricks, etc.

Try this technique for yourself by completing Hands-On Activity 3.1.

3.3 When and How to Use Detail

No matter how much you simplify your world, there will be occasions where detail is important—or a *representation* of detail. Detail in this case does not need to follow the subject matter line for line, but nor can it be simplified as a number of outlines. In these cases you need to be relaxed; simple lines, squiggles, and dots can help the viewer "complete" the image.

As humans we are hardwired to "autocomplete" the world around us—for example, we know that if a cow stands behind a tree, the cow does not become cut in two; the rest of its body is occluded. Likewise, even the most sketchy of faces can be detected as a face, even if it is two dots, a line, and a circle for the head. Studies with newborn babies show we still prefer to gaze at these "faces" rather than random lines! (Valenza et al., 1996).

So how can we utilise this knowledge for our icon practice? Firstly, find an object you think would be hard to depict with simple lines alone, and sketch out the bare bones. In the case below (Fig. 3.3), it is not obvious what the original image is (a book, a pamphlet?)—but when we add the representation of the detail, it becomes more likely that this icon is in fact a UK passport. We haven't explicitly drawn a lion or unicorn, but there is a hint of a horn and curves that represent the animals. A simple block represents the biometric icon.

Of course, this is fine if you are expecting to see a UK passport, but not all of these documents are the same—you must consider your audience. If you are working elsewhere in the world, your "hints" to detail might be different.

Here's another example—notes and coins. Many countries have faces on their banknotes and a denomination; we can approximate these with relatively few simple lines. The same goes for bank or credit cards. You'll have noticed in these simple icons that we frequently use simple, straight, or lightly wiggled lines to represent lines of text. In the case of particular items, the *placement* of these lines is also important.

Practical Application Tips

- The secret to representing or hinting at these details is not to try too hard. Don't overthink it, just quickly sketch out an approximation of the layout, item type, and any defining shapes or logos—such as the bank card logo in the bottom right of Fig. 3.4.

  Try this technique for yourself by completing Hands-On Activity 3.2.

Fig. 3.3 Passport icon detail development, *Microsoft Paint* on *Microsoft Surface Go* using *Microsoft Surface Pen*. Miriam Sturdee, 2023

Fig. 3.4 Currency icon detail development. *Microsoft Paint* on *Microsoft Surface Go* using *Microsoft Surface Pen*. Miriam Sturdee, 2023

3.4 Combined Visual Icons

To begin to sketch the world, we need to first look at what surrounds us, the expansive seas, luscious forests, and bustling cities. They are made up of beings, objects, metaphors, and concepts that when separate may not hold much meaning but, when combined, can express experience and emotion. Remember, looking at the world line by line, simplifying (what can be kept and removed without losing its essence), and exploring non-tangible objects through sketching can help us to build not only the world through our fingertips but also allow us to communicate silently with each other. By allowing visual icons to communicate with each other like a stranger approaching another to establish directions to the local supermarket, one can express a more complex visual dialogue.

For example, a visual icon of a toilet tells the viewer that they have reach the intended location; whereas if one were to combine this with a wheelchair, it becomes an accessible toilet; if a warning icon is placed over it, it is now out of order; or if an arrow is placed next to it, it becomes directions to the next nearest toilet. Here are a few examples (Fig. 3.5).

Practical Application Tips

- Practice by drawing from real objects whenever possible.
- Truly observe what you are sketching, often referred to as observational drawing, knowing what to keep and remove whilst maintaining the object's essence can only be achieved if you know what the object looks like in the real world.

Try this technique for yourself by completing Hands-On Activity 3.3.

3.5 Domain Visual Icons

Icons and images can be domain specific, but of course there will be times when they may overlap, be added to, or edited to make them more universal. An important thing to remember is that your audience will probably dictate what works and what does not. Take a look at the sustainability library (Fig. 3.8) as an example—there is an icon of an ice cream in a cone at the bottom right. At face value, this is simply a

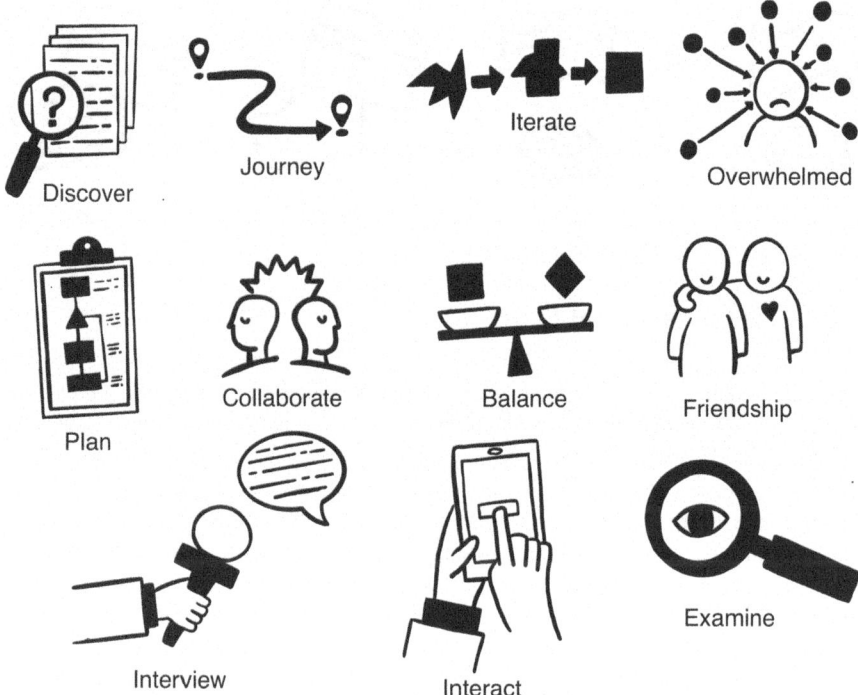

Fig. 3.5 Combined icons. *Photoshop* on *Microsoft Surface Pro* using *Microsoft Surface Pen*. Makayla Lewis, 2021

picture of an ice cream, but to the sustainability in computing community, this was developed to represent a niche concept: "vanilla sustainability". Likewise the crossed swords in Fig. 3.6 might mean "fight" to a general audience—but within the human-centred AI and law scholars, this takes on the meaning of "protection".

For the example icon libraries we have provided (Figs. 3.6, 3.7 and 3.8), try to guess what each icon might mean. If you are in a group, compare answers between yourselves. The variety may surprise you! This should show you how important *domain* and *context* are for interpretation.

Your situation and interests will be unique. It is likely you will have an overarching domain—perhaps your module or course, then a subdomain, perhaps some coursework or an assignment. Imagine you are working on a visual library for the user experience of a smart fridge, for example. Where might you start?

Well, a good place to start is the title and concept itself. You could start by representing "user experience" and "smart fridge". How you narrow down your domain will depend on your approach and role in your team or group. You may want to consider the interface, the user, food, drinks, temperature, and so forth. A successful domain-specific library of icons will help you develop stories, communicate concepts, and can even help design your interfaces and engage with paper prototyping.

3.5 Domain Visual Icons

Fig. 3.6 Human-centred artificial intelligence in law icon library. *Photoshop* on *Microsoft Surface Pro* using *Microsoft Surface Pen*. Makayla Lewis, 2021

Fig. 3.7 Human-centred cybersecurity icon library. Fineliner pen on paper. Miriam Sturdee, 2021

Fig. 3.8 Sustainability in computing icon library. Fineliner pen on paper. Miriam Sturdee, 2018

Practical Application Tips

- Brainstorm your domain in words first to map out the possible items you will want to represent.
- Use these words to compare and discuss with your peers to identify gaps.
- Refine the essential items for the domain-specific library.

Build your own domain-specific library by completing Hands-On Activity 3.4.

3.6 Hands-On Activities

Activity 3.1: Simplify Your Environment (Individual)

Learning objective—To observe, understand, and apply the idea of building an icon library based on your surroundings

Time—15 minutes

Materials—10 to 20 Post-it notes, 1 black pen, pencil, and your place of work (workstation, desk, or table)

Procedure:

- Looking around your chosen location, identify ten unique objects, e.g. a lamp, book, laptop, plant, glass of water, headphones, etc.
- Observe the object from various angles considering what is unique about the object:

  (a) What is common about the object (e.g. pens usually have caps/lids, a nib; a mug has a handle and holds liquid)?

  (b) What can be omitted?

  (c) What can be enhanced?

- Using a Post-it note, sketch the object clearly, concisely, and quickly. Using the skills you learnt from this chapter, draw an icon that represents the object.

3.6 Hands-On Activities

- Stick your completed Post-it note to the chosen object.
- Go to your next object and repeat the previous two steps.
- Once you have completed your icons for all ten objects, return to the first object.
- Take a step back; review your icon for clarity and accuracy:
  - (a) If you think your icon requires adjustment, rather than draw over it, grab another Post-it note, and try again.
  - (b) If you think your icon truly reflects your object, move on to the next object.
- Repeat the previous step until you have reviewed all icons.
- Gather your Post-it notes and stick them in your sketchbook. Do not forget to write the object name underneath your icon. You have now started to build your own icon library (Figs. 3.9, 3.10, 3.11, 3.12, 3.13 and 3.14).

Activity 3.2: Representing Details (Individual)
Learning objective—To refine and develop icons or imagery where detail may be necessary to distinguish from similar objects
Time—10 minutes
Materials—Sketchbook or paper, 1 black pen, a selection of objects from your desk or workspace with varying levels of detail, your icons from Activity 3.1
Procedure:

- Look around your environment for items that have little details that make them unique. These could be famous album covers or logos, a particular tv show to place inside your "smart-tv" icon, your favourite app for your smartphone, or a jar of jam.
- Sketch three of these items quickly, using the basics you learned from Activity 3.1, and then quickly add hints as to the detail—don't get bogged down in being precise; this should be a quick squiggle or shape which is recognisable without being fussy.

Fig. 3.9 Real fire extinguisher (left) and fire extinguisher icon (right). Photograph and *Procreate* App on *Apple iPad Pro* using *Apple Pencil*. Makayla Lewis, 2021

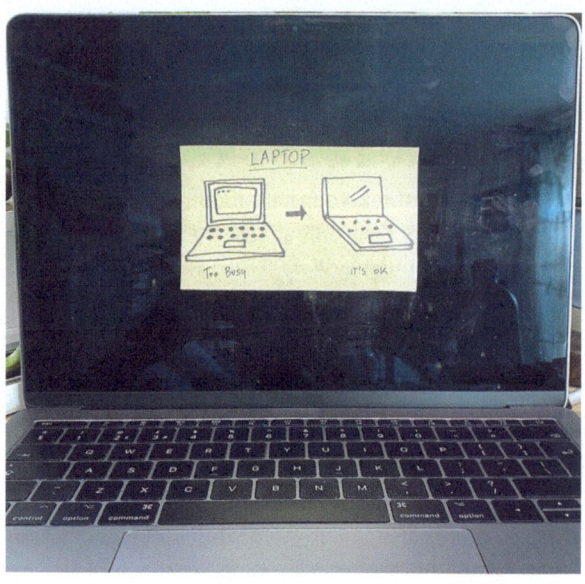

Fig. 3.10 Evolution of a laptop icon. Photograph and fineliner pen on Post-it note. Makayla Lewis, 2021

Fig. 3.11 Icons from Miriam's current surroundings at the time of writing, *Procreate* App on *Apple iPad Pro* using *Apple Pencil*. Miriam Sturdee

3.6 Hands-On Activities

Fig. 3.12 Living room (left) and kitchen (right) environment and icons. Colour pencil on paper. Makayla Lewis 2020

- For each item, make three attempts. Try different approaches to each item—for example, outlines and dots, or squiggles and cross-hatching—which approach is more recognisable?
- Compare these images to the basic icons from Activity 3.1; do they still appear to be in the same style you started with?
- Collect your icons together to revisit later; they will be useful for helping develop your ideas for Activity 3.3.

Activity 3.3: Rapid Word Visualisation (Individual)
Learning outcomes—To quickly apply the idea of visual icon building based on predefined words
Time—20 minutes
Materials—Sketchbook or paper, and 1 black pen
Procedure:

- Using the grid below or a separate piece of paper.
- Start a timer for 30 seconds; for the first ten sections in your mind, visualise the word; the preceding ten sections, consider how you could simplify the word and then rapidly sketch the word, 10 seconds maximum, using the learnings and tips from Sec. 3.3.
- The first four words have been completed as an example (Fig. 3.15).

Fig. 3.13 Bedroom (left) and coffee shop (right) environment and icons. Colour pencil on paper, Makayla Lewis, 2020

Fig. 3.14 Representing details of a calculator. *Procreate* App on *Apple iPad Pro* using *Apple Pencil*. Makayla Lewis 2023

3.6.1 Visual Icon Grid Activity Sheet

Activity 3.4: Domain Icons and Rapid Icon Generation (Group Activity)
Learning objective—To be able to identify concepts, brainstorm, and sketch appropriate domain-specific icons
Time—20–30 minutes

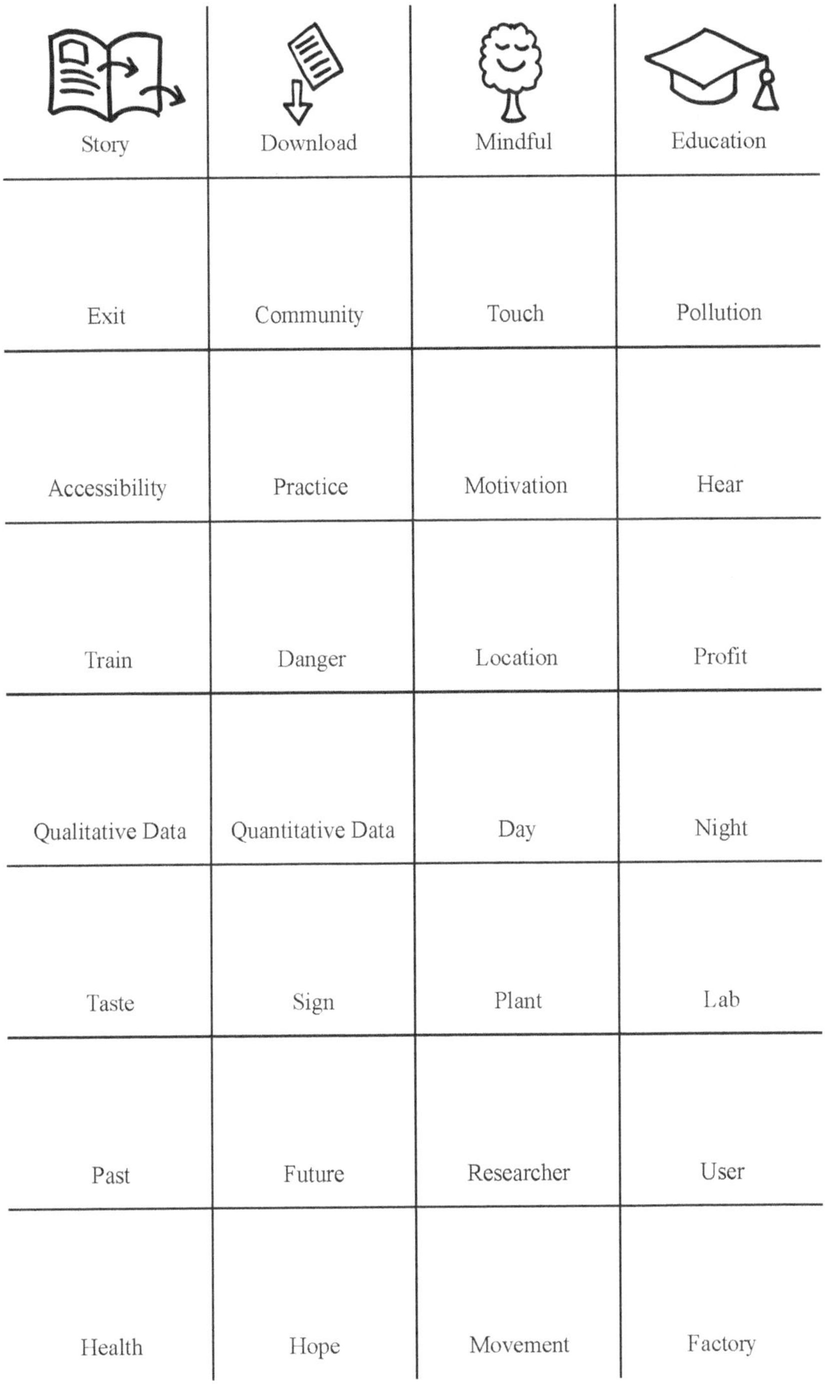

3.6 Hands-On Activities

Story	Download	Mindful	Education
Exit	Community	Touch	Pollution
Accessibility	Practice	Motivation	Hear
Train	Danger	Location	Profit
Qualitative Data	Quantitative Data	Day	Night
Taste	Sign	Plant	Lab
Past	Future	Researcher	User
Health	Hope	Movement	Factory

Fig. 3.15 Visual icon grid activity sheet

Materials—20 to 30 Post-it notes, 1 black pen, 1 red pen, a large wall, window, or whiteboard to stick Post-it notes on

Procedure:

- Identify your (or your group's) core area of interest in UX or HCI.
- By yourself, or as a group, write down key concepts, items, or ideas central to that theme—each one on a separate Post-it note. Try to generate at least 20 title concepts.
- Stick these Post-it notes at the top of your wall space, window, or white board. Not too high—you want to be able to create a "drop-down" of Post-it notes underneath that everyone can reach.
- For *each item* everyone taking part in the activity should rapidly sketch their interpretation of the icon. If you are by yourself, try 3–4 different iterations for each.
- When you have finished sketching, go along the list of Post-it titles, and stick all the variants underneath in a long line.
- It's voting time! Take out your red pen, go along the line, and put a small red dot in the bottom corner of your favourite icon for each concept. If you are in a group, count the votes and see which icon has the most votes.
- Discuss (if in a group) and think about why you like the favourite icons and how each one could then be refined or improved.
- Each "favourite" should then be refined and redrawn neatly to create the overarching icon for that concept. These can then be collated, scanned, and digitised to create a visual icon library reference for your domain (Figs. 3.16 and 3.17).

Fig. 3.16 Exemplar user experience domain icons. *Photoshop* on *Microsoft Surface Pro* using *Microsoft Surface Pen*. Makayla Lewis 2019

Fig. 3.17 Monthly icon calendar created for *Sketchnotes* Community. Fineliner pen and marker on paper (Template: *Adobe Photoshop* on *Cintiq Companion*). Makayla Lewis 2020

References

Books, Papers, and Articles

Neurath, O., Eve, M., & Burke, C. (2010). *From hieroglyphics to Isotype: A visual autobiography*. Hyphen Press.

Valenza, E., Simion, F., Cassia, V. M., & Umiltà, C. (1996). Face preference at birth. *Journal of experimental psychology: Human Perception and Performance, 22*(4), 892.

Websites

W1 Helpful icon library, use with attribution or buy a subscription: www.thenounproject.com/

Further Reading

Brand, W. (2018). *Visual doing workbook*. Laurence King Publishing.
Brand, W. (2021a). *My icon library: Build & expand your own visual vocabulary*. BIS Publishers.
Brand, W. (2021b). *My icon library: Build & expand your own visual vocabulary*. Laurence King Publishing.
Hall, R. (1997). *The Cartoonist's workbook drawing* (1st ed.). A & C Black Publishers Ltd.
Miyata, C. (2016). *How to draw almost everything: An illustrated sourcebook*. Quarry Books.
Rohde, M. (2013). *The sketchnote handbook: The illustrated guide to visual notetaking*. Peachpit Press.
Toselli, M. (2016). *100+1 drawing ideas for sketchnoters and doodlers*. Independently Published.

Chapter 4
Text, Connections, and Colour

4.1 Describing Your Visual World

Hopefully you've built up an impressive range of icons and are ready to start thinking about how to connect these sketches to each other and describe them in a way that helps tell a story, interrogates, or communicate a concept. Don't stop drawing icons however—if you come across new objects, at study, work, or play, add to your knowledge base. If you're keeping a sketchbook, then try keeping it to hand, and if you find you have some downtime at a coffee shop, or waiting for an appointment, look around you, and add some icons to your visual vocabulary.

One of the good things about icons is that they communicate an object or concept succinctly and clearly, but they do not usually exist alone on the page. This chapter will help you think about constructing a page with text, connectors, and colour, simple techniques to direct the viewer between points of interest and elaborate upon areas of focus.

Although we obviously advocate communicating visually with sketches, it is rare that a complex concept or page of multiple sketches can be fully described with the drawn line alone. If you are going to use your sketches as a form of communication, or use them in your writing and research or studies, then it helps to title, annotate, and direct the viewer to each relevant section in turn. To do this we necessarily use text, connectors, and colour to curate the gaze.

By the end of this chapter, you should be able to:

1. Use text confidently in different weights and styles.
2. Connect and separate sketches on a page in a variety of creative ways.
3. Develop a colour palette for your sketches that works for you and your projects.

4.2 Using Text in Sketches

You are probably more familiar with handwriting than sketching! But remember when we told you in Chap. 1 that drawing and language develop along similar pathways in the brain (Cohn, 2012)? Now is the time to unite those skills.

We aren't going to teach you how to write (hopefully you have already learned that!), but we will give you ideas on how to work with your own handwriting and develop strategies for different uses of text within your sketches. If you already have a neat and tidy handwriting style, either *cursive* (joined up) or *cuneive* (individual letters), then you can enjoy lettering the main body of your sketched page as you would writing a letter.

First things first, if you are constructing an informative page of sketches, you might want to give it a title, such as "Research Process" for example, if you are using your sketching skills to envisage your project's workflow. The title will of course depend on your interests, studies, and research.

Titles are, by necessity, bigger and bolder and more visible than general annotation or labelling text. A title should draw the eye and inform the viewer of what to expect in the sketched page. In this context, it is a good idea to move away from your "general" quick handwriting that you use for writing notes—keep that for later. There are a few ways to deal with bigger text and titling, and you should also consider if you want to use subtitles as this will affect your "top-level" title.

The text does not need to be complicated or elaborate; you only need one style that can be easily adjusted to show text priority, i.e. adding additional lines and changing weight can be quick and incredibly effective (see Figs. 4.1-1)—there are four initial lettering styles. From there, your "letters are your oyster". Following

Fig. 4.1 A range of approaches to text, from different weights to showing concepts with text. (1) Four core lettering styles, (2) adjusting core lettering styles, (3) lettering guides to support less wonky lettering, (4) examples of using lettering guides. *Procreate* App on *Apple iPad Pro* using *Apple Pencil*. Makayla Lewis

this, you can be more creative by making minor adjustments, e.g. replacing lines with dots, accentuating elements of letters, or omitting a few less essential features of a letter (see Figs. 4.1-2). If you are like us, trying to write along a straight line unaided can lead to pages that appear "wonky"—often slanted to the right. We found that this is one of the rare moments when a pencil can help you return to primary school; the guides you were given to ensure letters were written correctly can be helpful (see Figs. 4.1-3). However, it is rather time-consuming to create; we recommend using a single line you manipulate and then writing on top of it. Once you complete your letters, grab the eraser, and remove the pencil line; no one will know (see Figs. 4.1-4).

For further textual study to support your burgeoning sketching skills, you may also wish to experiment with calligraphy (Heller & Talarico, 2011; Seddon, 2013) and purchase a calligraphy pen. You can buy a range of chisel-nib calligraphy brush-markers in different sizes from many major brands, or if you wish to truly engage with lettering, then investing in fountain pens, dip pens, and Chinese or Japanese calligraphy brushes is a must. However, this book is about sketching in HCI, so we must let you embark on that journey with other support! It is possible, of course, to use calligraphy pens and brushes for sketching though, and we encourage you to experiment.

Writing directly in "bubble" or outline text is hard. If you look at the "bold bubble" in Fig. 4.2, it looks easy, but you can trip up over some letters, like lower case "e" and "S". Practice will help hone these skills, but there is a nice trick if you have a thicker black pen with a round nib—simply write the text with that pen, and tidy up with your fineliner. This is even better if you use a coloured pen, as you then have a bright and bold title, which you can also then outline using your fineliner—instant title appeal! (Fig. 4.3)

Fig. 4.2 Three more styles of text which work well for titles, drop-shadow bold bubble, side-shadow single stroke, and accent-shadow single stroke. *Procreate* App on *Apple iPad Pro* using *Apple Pencil*. Makayla Lewis, 2019

Fig. 4.3 Easy, quick bubble text for titles, leave as is, or outline for emphasis. *Procreate* App on *Apple iPad Pro* using *Apple Pencil*. Miriam Sturdee

If you want to spend a little more time on titles, for example, you are handmaking a poster, zine, or other sketched piece that requires a bit more "zing", then you can take things a little further. For fun and illustrative titles, you can turn to your favourite word processing program and explore the fonts they have listed, and use them as a resource, or find fonts online from free services for inspiration. Company logos can also help you connect icons and titles at the same time and give you ideas for colour schemes *within* your text—just be careful not to breach anyone's copyright (the exception being if you are creating a piece *for* a company and wish to copy their logo freehand). The logos in Fig. 4.4 are from start-ups based in Lancaster, UK, and were sketched with permission as part of an event to get students interested in entrepreneurship and innovation.

As well as logos providing inspiration, your textual sketching skills may also help you to design logos and imagery for your own projects, start-ups, or future ideas. Text and iconography together can be a powerful tool to communicate meaning and background. Activity 4.5 in this chapter suggests ways in which you can practice this. In Fig. 4.5 you can see some examples of adding icons, but also borders and combinations of styles—elaborating upon your text can make it really stand out, or even make it part of the story of a page.

You are probably familiar with "thought bubbles" and "speech bubbles" from comics, and square borders and underlines—but there are many different styles you could adopt and make your own, from simply doubling up a rectangular border into two lines to adding a drop shadow or even a scroll (Fig. 4.5, bottom left). In some examples there are also fancy borders with imagery, icons, and people! Not confident in drawing people? Don't worry—we cover that in Chap. 5 (People, Faces, and Actions).

For those who prefer to push their lettering further, brush lettering is for you. It combines a pen or marker with a brush tip, unlike calligraphy, which you have probably heard of (and if you are like us), you fear it. Brush lettering is an artistic form

Fig. 4.4 Sketchy logos and elaborated text, using text, shape, and colour for emphasis, fineliner pen and marker on Bristol board. Miriam Sturdee, 2019

4.2 Using Text in Sketches

Fig. 4.5 Sometimes imagery can help a concept "pop!"—top left you can see a match under the "Ignite 100" text. This is not a logo but illustrates the semantic meaning of the word. The same can be seen with the axe for "hacks". Other icons are more explicit, for example, "rent a textbook" has a simple book icon. Fineliner pen and marker on Bristol board. Miriam Sturdee, 2019

Fig. 4.6 Example of incorporating brush lettering into your sketches (and sketchnotes)—*Sketchnote LDN* of "Building a visual community at lernOS Sketchnoting Guide (English Version)" W3 release party. Fineliner pen and marker on paper. Makayla Lewis, 2020

of lettering that is the easier best friend of calligraphy; it uses a similar method of alternative pressure to vary the line weight but can be learned quickly. Thus, including it in your sketches can help you add interest (Fig. 4.6); therefore, it is often used for titles or to draw your viewer's attention to a word or quote from a participant/user.

We recommend using Tombow dual brush pens or Pentel Fude pens to practice. Using a grid, as discussed previously, hold the pen or marker with a controlled grip, ensuring your movement is fluid:

- Experiment with pressure; lift gently (upstrokes). You will see the line weight decrease; press hard (downstrokes), and it will have the opposite effect.
- Plan your letters and grab a pencil to help you with this.
- Start with the basics, the alphabet, e.g. Fig. 4.7.
- Then work on connecting the letters, e.g. Fig. 4.8.
- Practice to ensure consistency in spacing, size, and style, thus providing improved legibility for the viewer.

Do not worry; you can practice grids and letters; and also do not worry if you make a mistake; it's okay; it's a technique that takes time to master.

Practical Application Tips

- Remember you do not have to change HOW you write; just jazz it up, and work to your strengths.
- Keep titles short whilst you are practicing; it is easier to be consistent in small chunks.

Fig. 4.7 Example of incorporating brush lettering into your sketches. *Procreate* App on *Apple iPad Pro* using *Apple Pencil*. Makayla Lewis

4.2 Using Text in Sketches

Fig. 4.8 Example of connecting letters. *Procreate* App on *Apple iPad Pro* using *Apple Pencil*. Makayla Lewis

- Keep a visual record of fonts and logos you see around you, sketch them out quickly, and find styles you like.
- Work to scale—choose large, thick markers and pens if you are using larger paper sizes—a title in 0.5 mm width fineliner pen will be lost on an A4 and A3 page.
- If you struggle with keeping your text on track, try practicing on squared or lined paper.
- Keep writing! We spend so much time typing these days that handwriting is becoming a lost art. Like sketching, handwriting can improve with practice. Try writing yourself little motivational notes—"I must keep sketching".

Try this technique for yourself by completing Hands-On Activity 4.1.

4.3 Simple Connections

Drawing a line between two objects is the easiest way to show a connection, and you've probably used this before in diagrams or mind maps—and you could just leave it there! However, by utilising simple arrows or other end points, you have more control over the link and can use it to communicate a personal style, branding, or colour-code for more complex information.

If you look at the example (Fig. 4.9), we take the humble line and elevate it. It can be made thicker, given arrow heads of different varieties; it can even turn into footprints! In two examples the line connects boxes or bubbles which also contain objects. You could also use a different pen, a colour, or even a simple highlighter marker to connect your elements. A connection could also be simply a colour that is mirrored elsewhere on the page—linked conceptually rather than with a direct line. We talk more about highlighting and consistency in colour use in Chap. 6.

"Linked" to the idea of connections is that of separators! Sometimes we want to divide our page or objects on the page to ensure that they don't "spill over" into the wrong place. The simplest way to do this is to "box things off"—a grid is the simplest way to divide a page—but you don't have to be square about it! You could draw lots of circles, or bubbles, or simple lines or zigzags between sections to show the distance between concepts. Figure 4.9 shows some simple separators for you to try—but you'll likely discover more as you go along. Figures 4.10 and 4.11 show separation in practice whilst sketchnoting, both in analogue and digital contexts.

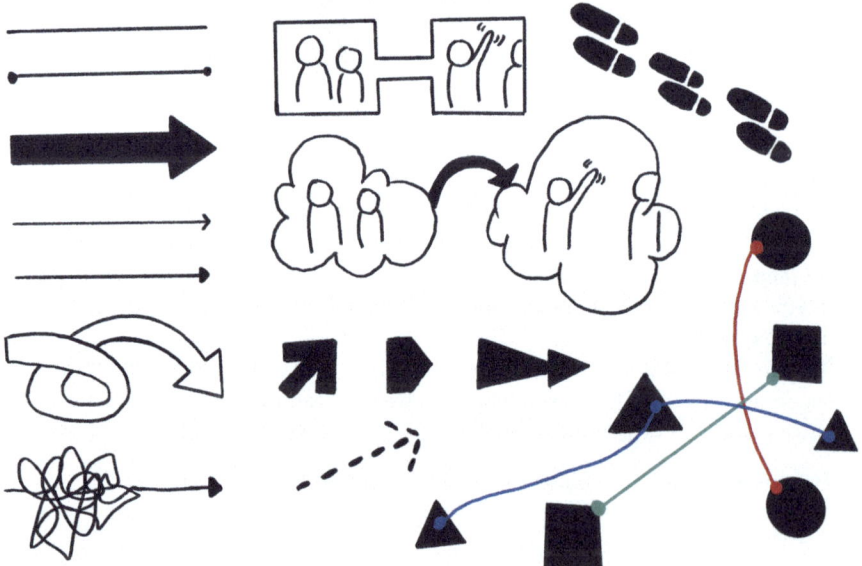

Fig. 4.9 Simple connections, with a variety of end points to the line; note the use of thickness of line and colour for emphasis. *Procreate* App on *Apple iPad Pro* using *Apple Pencil*, Makayla Lewis

4.3 Simple Connections 61

Fig. 4.10 Separating information can be as important as connecting it! Being clear with layout and items will improve the communication of your sketches. Sketchnotes from *Sketchnote LDN* visit to British Museum. Fineliner pen and marker on paper. Makayla Lewis, 2019

Fig. 4.11 Sketchnotes from ACM the Future of Computing and Food workshop. *Photoshop* on *Microsoft Surface Pro* using *Microsoft Surface Pen*. Makayla Lewis, 2018

Practical Application Tips

- Not everything needs connecting or separating; try to keep your page uncluttered—sometimes white space speaks volumes!
- Try different weights of fineliner pen, and in different marker colours; be consistent within a page or project.
- To draw attention to a particular item, double up or thicken lines, or use colour only in that one place.
- Branch out from squares or circles as separators, and try tesselating shapes like triangles, frames, or hexagons for a fun visual approach.
- Don't overthink or rush connectors and separators; they can be added later when you have time.

Try this technique for yourself by completing Hands-On Activity 4.3.

4.4 Colour Your World

Black and white, monochrome, and greys can be highly communicative, and often you'll be quickly sketching out a concept with your pen and not have time to think about elaborating upon it, save for some quick connectors or separators perhaps! However, even with one other colour than black, you can quickly help communicate key points, help icons or imagery stand out, and show how sketches relate to each other. You can also sketch icons, for example, directly in colour, and if you have more time and more colours, you can also create highly elaborate and communicative sketches with many colours—but beware! Too much colour in a simple sketch can be overwhelming.

We learn about colour from all sorts of sources in school: colour from the arts, what colours mix to form others (Fig. 4.7); in computing, how colour translates between digital and print; from biology, how we perceive colour; and from physics, how light reflects and combines to produce colours (Mikellides, 2012). These are basic examples, but we don't need to go into a lot of depth here about these approaches, because sketching with colour is a much freer and easier experience (Figs. 4.12, 4.13 and 4.14).

The easiest way to get started with colour is to pick one that you like and buy a dual-tip marker that allows you to add fine detail and connections but also block out or shade areas if needed. If you are a huge fan of dark colours such as navy though, then be aware that they may not contrast enough for people to discern the difference. Colour contrast, hue, and saturation are also an important aspect of the accessibility of your sketches—something we cover in depth in Chap. 8. Usually choosing a light grey if you want to do some shading or shadows and one other colour should be sufficient. If you are going to be sketching for the lifetime of a project however, you may want to expand your colour selection into a palette and use it consistently to signpost the viewer and add visual appeal (try Activity 4.4 to explore making a colour palette).

Fig. 4.12 Colour theory can help with the basics, but often you might want a more subtle contrast between colours, or get a palette for a particular project. 1. Colour palette, 2. hues 3. greyscale 4. saturation. *Procreate* App on *Apple iPad Pro* using *Apple Pencil*, Makayla Lewis

Practical Application Tips

- Less is often more; have a "go-to" selection of coloured pens that you carry with you—it also saves time.
- Make sure your colours don't get "lost"—choose colours that have a reasonable contrast so you can use them throughout your sketched page, and together.
- Save your brightest and most eye-catching colours for the most important thing on the page!
- Develop different palettes for different projects; the visual change will help keep things focused.
- Make colour swatches when you get new pens, so you know how the colour appears on the page—also try photographing or scanning colours—sometimes the pigment does not translate well to digital contexts.

Try this technique for yourself by completing Hands-On Activity 4.4.

4.5 Hands-On Activities

Now try the activities below!

Activity 4.1: Playing with Textual Style (Individual)
Learning objective—Try out the styles illustrated in the chapter. Gain confidence in adapting your own handwriting and working with title text
Time—20 minutes
Materials—Black fineliner, coloured marker with a chunky tip, calligraphy/chisel tip marker (optional), plain paper or sketchbook

Fig. 4.13 Defining a colour palette for your sketching can help bring the image together and show cohesion between elements. Here, this sketchnote uses grey and yellow. Fineliner pen and marker on paper. Miriam Sturdee, 2016

4.5 Hands-On Activities

Fig. 4.14 Defining greyscale for your sketching can help bring the image together and show cohesion between elements. TodaysDoodle No. 858—sketchbook page from *Gosh!* Comics "Drink, Comedy and Draw online" session. Fineliner pen and marker on paper. Makayla Lewis, 2020

Procedure:

- Think of a short sentence, for example, "The rabbit is sitting on the mat".
- Write this sentence in your own handwriting.
- Look at your handwriting, is it legible? If not, write the sentence again, slowing down and paying attention to detail.
- Write the same sentence in capital letters. Again, look at your preferred handwriting style. Do you have an interesting way of drawing the letter "A", for example? "Capitalise" on the things that make your handwriting unique!
- Now try some title text. Choose something short so it doesn't take too long to write it out several times, such as "Rabbit Habits". Try out the different line styles from Figs. 4.1 and 4.2, then get your marker, and try the bubble text from Fig. 4.3.
- Choose your favourite title text style and elaborate upon it, with an icon or some shading.
- Keep practicing, and keep your examples as a record of progress as you work through the rest of the book.

Despite having written essays by hand at school and using handwriting on my sketches, I still wish I (Miriam) wrote in a beautiful calligraphic style or had the sort of handwriting that elevates a greeting card! However, by using the tips and tricks in this chapter, I find I can create titles I am pleased with and have also learned that my capitals are much easier to read than my normal, fast handwriting.

Activity 4.2: Font Finding (Individual or Pair Activity)

Learning objective—Learn how to emulate different text styles, and gain confidence in applying different styles

Time—15–20 minutes

Materials—Black fineliner, thicker black marker pen, paper or sketchbook, a computer or tablet

Procedure:

- On your computer, open either a text editor or web browser—if you have no computer, find some printed material nearby with different fonts and title weights.
- If using a text editor, write a short sentence into a new document, and then select an interesting font of your choice from the font selection tool (e.g. you could choose something with fancy filigree lines or chunky, bold appeal). Apply it to your sentence.
- If using a web browser, try searching for reputable font providers, such as fonts.google.com, and pick a fun font, and then type some sample text. Keep the results on screen.
- If you are working in a pair, show your partner the example, and allow them to have it in front of them to work on.
- You should now have a short sentence in an interesting font. Examine it. Is it serif or sans serif? (Serif refers to the little "feet" or additions at the end of the letter lines; sans means without—e.g. Arial). Is it curvy or straight? Does it have a unique aspect to it? Is it pictorial in nature?
- Imagine that each letter is an icon or object to be sketched. Start with the basic form and lines, and then add weight and any distinct features.
- If the font is thick, try using the black marker for the basic lines, and then use the fineliner to create crisp edges and details.
- Return to the beginning of the activity, and now choose the most outrageous and fun font you can find. Again, swap with your partner if you are working in a pair, or start exploring the form by yourself.
- This is a good thing to practice in downtime or during meetings or class to keep your hands busy. You can build up a "repertoire" of styles and use different text in different situations—quick blocky titles for when you don't have much time and delicate, winding lettering to add character to a slow and detailed sketch.

These textual styles will come in useful in Chap. 6 when we start bringing different sketched elements together.

Activity 4.3: Divide and Conquer! (Group Activity)
Learning objective—How to connect and separate sketches in a variety of ways
Time—30 minutes
Materials—Black fineliner, chunky black pen, a different colour medium-nib marker for each group member. A piece of A4 paper or sketchbook page for each group member. Two or three pieces of A2/flipchart paper, neutral-coloured or white sticky notes, or cut out squares of paper. Work around a table.
Procedure:

- Each group member takes two squares of paper or two sticky notes. Using a black fineliner, each person should then draw two icons of their choosing from their favourites in Chap. 3.
- Once the icons have been drawn, they should all be placed on the large-format paper in the middle of the table. Space them all out evenly.
- Starting with the person with the lightest colour (e.g. yellow) connect two of the icons with a chosen style of connecting.
- Next, the second lightest colour owner should try to connect another two icons, without sketching OVER the previous person's connectors, but cross-over is allowed. Each person should try and think of a new way to do a connector.
- Each member of the group should have a turn, until nobody can think of a new way to connect the icons.
- As a group, discuss your favourite and least favourite connectors, when might you use different styles? Together, use either the chunky black pen or fineliner to highlight and edit them.
- Each group member should record their favourites, using either the black chunky pen or fineliner.
- Repeat the exercise for separators, using a new sheet of A3 paper if the page gets too busy. Except each group member should separate one icon from another, using a line or full bounding box.

Activity 4.4: Colour Me Beautiful (Individual Activity)
Learning objective—Build and use a colour palette; utilise it in icon work, connection/separation, and text
Time—10 minutes
Materials—All the coloured pens and pencils you have to hand, a piece of paper, a black fineliner
Procedure:

- Pick your favourite colour.
- Look around your space and find objects with different hues of that colour.
- Bring them together and take a photo.
- Grab colour pencils or markers, recreate the image in icon form, and then colour each item in the correct hue.
- If you do not have the exact colour, try mixing them.

Activity 4.5: Rapid Icon Storytelling (Pair Activity)
Learning objective—You will be able to create a simple icon story
Time—5 minutes
Materials—All the coloured pens and pencils you have to hand, a piece of paper, a black fineliner
Procedure:

- Using the icons you created in Chap. 3, bring them together to tell a story.
- Randomly number each icon.
- Decide on the type of connectors you will use.
- For each icon (in the order in which they were numbered), create a visual event, e.g. if your icon is a loaf of bread, you could draw someone next to the loaf in the kitchen, cutting.
- Ensure each icon follows on from the previous so it is a continuous story—get creative; no one says it needs to be a serious or realistic visual story; the fantastical, the better.
- If you have an object, stick figure, or scene that looks "questionable" (no one would understand it), add a callout arrow and describe your sketch in your best new textual style. Don't worry we usually draw objects, people, or a scene that Makayla usually calls "funky" that requires extra explanation.

Please note storytelling and narratives will be explored in greater detail in Chap. 6.

References

Books, Papers, and Articles

Cohn, N. (2012). Explaining 'I can't draw': Parallels between the structure and development of language and drawing. *Human Development, 55*(4), 167–192.
Heller, S., & Talarico, L. (2011). *Typography sketchbooks* (p. 368). Thames & Hudson.
Mikellides, B. (2012). Colour psychology: The emotional effects of colour perception. In *Colour design* (pp. 105–128). Woodhead Publishing.
Seddon, T. (2013). *Draw your own fonts: 30 Alphabets to scribble, sketch and make your own*. Princeton Architectural Press.

Websites

W1 Some useful tips for choosing fonts and styles of text – www.fonts.google.com/knowledge/choosing_type/emotive_considerations_for_choosing_typefaces
W2 Activities for improving your lettering – www.makaylalewis.co.uk/category/blog/
W3 LernOS Sketchnoting Guide – https://cogneon.github.io/lernos-sketchnoting/en/0300_Introduction/

Further Reading

Heller, S. (2012). *Typography sketchbooks*. Princeton Architectural Press.
Lidwell, W., Holden, K., & Butler, J. (2010). *Universal principles of design, revised and updated: 125 ways to enhance usability, influence perception, increase appeal, make better design decisions, and teach through design*. Rockport Pub.
Macmillan, N. (2006). *An AZ of type designers*. Yale University Press.
McCandless, D. (2012). *Information is beautiful* (p. 978-0007294664). Collins.
Tselentis, J. (2011). *Type form & function: A handbook on the fundamentals of typography*. Rockport Publishers.

Chapter 5
People, Faces, and Actions

5.1 Beyond the Stick Person

Sometimes when we ask others to draw, we get asked "Are stick people ok? I can't draw..." but sketching a stick person *IS* perfectly acceptable (see XKCD.com – W1) and can be a great precursor to sketching *other* types of people. If you add legs, arms, and a head, what you have created will be seen as a person, regardless if it is blob-like. Children's sketches are wonderful measures of how shape and line can create the appearance of a person.

Drawing people is often the hardest thing to convince people to sketch in a class; the fear of "getting it wrong" can produce a creative block. But getting it "wrong" can actually be quite fun! Have a go at Activity 5.1 to get started by "sketching your neighbour". If you are working on your own, you could also use a mirror however!

Sketching for *Human*-Computer Interaction necessitates the inclusion of images of people, whether they are simply icons, interacting with computers, the world, or reacting to something. People express emotions, get frustrated, are happy, relieved, tired... If you have trouble imagining what these emotions feel like, use your own face as a reference, or look online.

The good news is that photorealistic portraits are *NOT* needed for sketching people in HCI. What is needed is a simple approach to adding figures and expressions into your work, and this is what you will learn in this chapter. And hands. We will also get you to sketch hands.

By the end of this chapter, you should be able to:

1. Sketch simple figures doing simple actions.
2. Add faces, features, and expressions to your figures.
3. Sketch hands interacting with objects.

5.2 The Simple Figure

So let's get to work! In Fig. 5.1 you can see four examples of simple figures, starting with the stick person and moving across to a more fleshed out version. If you can sketch out a stick person, then you can certainly expand upon that. Look at the second figure, this is a continuous line and can be roughly sketched as quickly as the simple stick person. We often call this a star person; like a starfish they have no hands or feet but instead are a simple shape with the head replacing the fifth point of the star. It doesn't matter that their body is long and their arms wide; it is still recognisable as a person—and this is what we are aiming for—remember IDEAS NOT ART! The third person we call a "blob" person. This one has a squared off oval for the main part, and simple lines for arms and legs, a good intermediary if you aren't comfortable with continuous line sketching. The "blob" could be almost any shape, or even a simple scrawled circle. Finally, we combine the styles by using the "blob" body, but adding star people arms and legs to the side. There! You can now start using people in your sketching work!

Practical Application Tips

- Get used to sketching figures in each of the styles; quickly sketch each one ten times in quick succession until it feels as natural as handwriting.
- Don't get fixated on faces whilst you practice bodies; simple faces can be added later if needed.
- Try exaggerating arms or legs for a different effect.
- Think about body types and body shapes; how could you adjust these figures to better represent a diverse group of people?

Fig. 5.1 Four figures. From left to right, the stick person, the star person, the blob person, and the combined result of each style. *Procreate* App on *Apple iPad Pro* using *Apple Pencil*. Makayla Lewis

5.3 About Face

Faces can bring a person to life, regardless of the approach you use to sketch out their body. In Fig. 5.1 we added lines and dots to sparsely represent features; simplicity can convey a lot of meaning. Interestingly, it has been shown that even newborn babies can recognise two dots and a line as a face (Mondloch et al., 1999)—so no need to get too detailed at this stage. Noses can be explored later.

Figure 5.2 is your passport to expression! This approach can really help you create meaningful faces with simple lines, which you can then elaborate upon later. By sketching out a grid of face shapes (or even squares) and adding two dots just north of the middle, you have already created nine faces. Next, add the lines shown to the top and left of the grid (straight, up-turned, down-turned)—these are your prompts—blue is "eyebrow" and red is "mouth". Now fill down and across. You have now drawn expressions ranging from indifferent to wicked! If you wish to add a nose, the simple straight line approach will work here as well. A straight vertical means the person is looking straight at you, diagonal left is looking left, and diagonal right is looking right.

The humble line is now your tool in creating meaningful, expressive faces. Once you are confident in this, try elaborating upon your faces. The first step might be to divide the eyebrow line in two and, the next, to add ears and hair. Ears do not have to be the complex shells and lobes of realism; simply add a small "cup" handle to the side of the head. They do not even have to seamlessly merge with the face shape.

Hair and headwear are an important part of the human experience. You can simply draw over the face grid people with a black pen, but once you are feeling you have the expressive technique down, try drawing the lower half of the face only, then exploring different hairstyles. If you are short of ideas, take a trip out, and sit in a coffee shop or park, and observe the people around you. Do they have long or short hair? What colour is it? Is it tied back? Do they have a hat? Further, do they have facial hair?

Fig. 5.2 The face grid. *Procreate* App on *Apple iPad Pro* using *Apple Pencil*. Makayla Lewis

In Fig. 5.2 Makayla demonstrates how a hairstyle can be created just using your black fineliner pen. You can also add a semblance of colour, light or dark, by adding lines in. Or simply add the outline and block hair in using another colour. Try looking in the mirror and at your own hair; how might you represent it?

The challenge you may find is if you are sketching a person with both facial hair and extravagant head hair, in which case relying on the face shape as a starting point can be difficult. In this case, the advice would be to sketch the eyes, nose, mouth, eyebrows (unless obscured), and craft the hair around the features. A particularly large beard might mean you cannot see a neck, in which case just sketch the shoulders coming out of the side! (Fig. 5.3)

When you feel you have got the hang of basic faces, hair, and expressions, then you can start considering how gaze and visual interaction might happen. Don't worry if this seems ungraspable at present, faces and angles do take a lot of practice. You will only get better if you keep going! Makayla's example in Fig. 5.4 shows how a simple oval and lines of focus can help you construct an angle. Don't be put off by the extra lines; look at the small face to the bottom right of each diagram. The only time you need to consider more than the oval at the start is when you are drawing a side view, in which case try to leave a nose gap! In the next chapter, these directional faces and gaze will be useful when you are bringing together your practice so far into visual narratives.

Are you feeling more confident? Let us consider bringing in some colour. Not the palettes you created in the last chapter, but thinking about skin tone. *Google Research (W2)* has a very useful website that has a skin tone continuum that you can see illustrated with basic faces in Fig. 5.5. You can find markers that correspond well to these colours in stores, but we realise it is not always possible to purchase several pens at once! If this is the case, then greyscale is your friend. Having a few grey tones means you can add shade and depth and even overlay greys to get the effect you need. In Fig. 5.6 we have simply converted the skin tone values to monochrome output to show an example, and in Fig. 5.7 the faces have been sketched directly in greyscale.

Practical Application Tips

- Remember when we mentioned economy of line from the chapter on icons? The same can apply to faces. Don't overwork them. Simple lines can convey a wealth of expression.
- If someone has a standout feature, try to incorporate it—perhaps they wear large, funky earrings? Or they have a fabulous hairstyle? Make it a focus! But at the same time…
- If you are sketching a person in real life and they are likely to see the picture, flattery is key!
- Simplify or omit facial features you do not want to draw attention to.
- If you are drawing a person with glasses, try to get the shape right, and then add a dot for the eye inside each frame.

5.3 About Face

Fig. 5.3 Hairstyles, headwear, and facial hair examples. *Photoshop* on *Microsoft Surface Pro* using *Microsoft Surface Pen*. Makayla Lewis, 2019

5.4 A Handy Guide

The nemesis of many an artist and nonartist alike, the hand is a complex appendage with multiple joints and details, all of which can be easy to misinterpret. At school, a particular favourite challenge of our art teachers was to make you draw your own

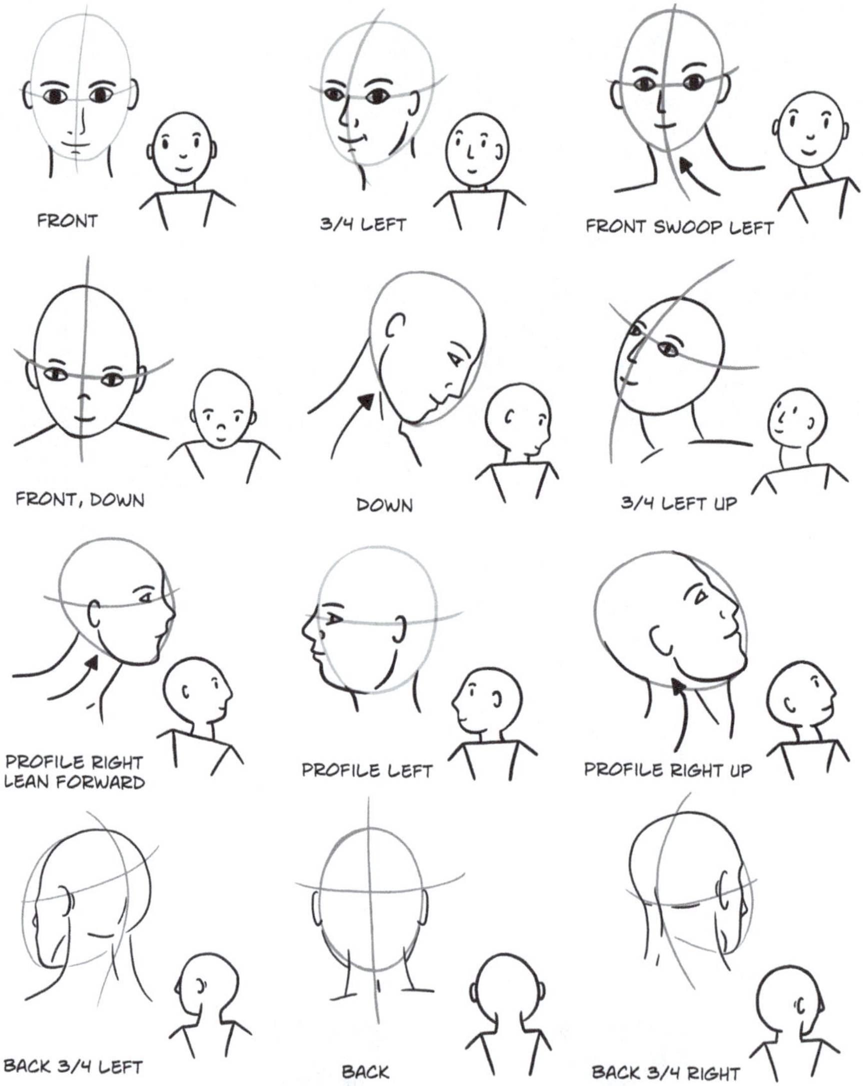

Fig. 5.4 Directional gaze facial construction. *Photoshop* on *Microsoft Surface Pro* using *Microsoft Surface Pen*. Makayla Lewis, 2019

Fig. 5.5 Monk skin tone, using suggested colour values from *Google Research* (W2). *Procreate* App on *Apple iPad Pro* using *Apple Pencil*. Makayla Lewis

5.4 A Handy Guide

Fig. 5.6 Greyscale skin tone. *Procreate* App on *Apple iPad Pro* using *Apple Pencil*. Makayla Lewis

Fig. 5.7 Example of using greyscale to represent skin tone. Makayla Lewis sketches using *Procreate* App on *Apple iPad Pro* using *Apple Pencil* (Lengyel et al., 2023)

hand. We still have nightmares about the deformed and awkwardly angled fingers and overly lined knuckles that resulted. Luckily, Makayla and Miriam have some simple simple tips and workarounds to help you out!

Much like the stick, star, blob, and hybrid people we looked at in Sect. 5.2, there are correlating hand styles you can add. In Fig. 5.8 you can see a variety of ways of appending a "stick arm" with shapes, and bottom left and top right show how you can either complete a star-person or hybrid arm with a simple point or scrawl five fingers or thumbs. When attached to a sketch of a figure, all of these will be recognisable as hands, and no time will be spent agonising over anatomy (Fig. 5.9).

Once you have found a style of hand you like, try appending it onto your figure style of choice! Figure 5.9 shows basic interactions using hands you may want to try, and you will utilise some form of hand in several of the activities at the end of this chapter. Of course, as hands are tactile appendages, you also need to consider how they might touch the objects around them, as well as gestures such as waving or pointing. In Fig. 5.10 Miriam shows both a pointing and pinching action, drawn

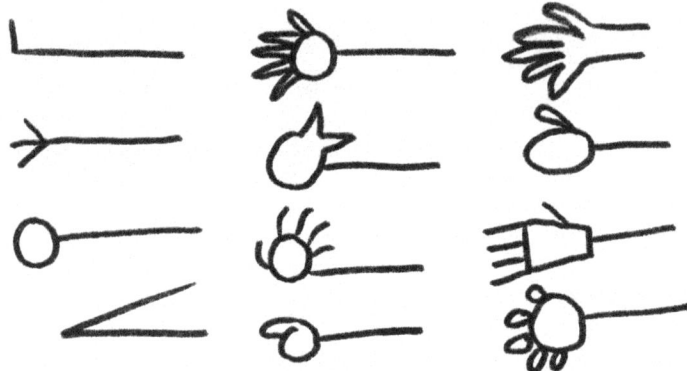

Fig. 5.8 At the end of the arm… a variety of ways to add hands to simple figures. *Procreate* App on *Apple iPad Pro* using *Apple Pencil*. Makayla Lewis

Fig. 5.9 Stick examples with hands—basic interaction with objects from previous chapter on icons. Fineliner pen on paper. Miriam Sturdee

from life (nondominant hand) and then scanned, coloured, and copied to save time. They are slightly wonky, and maybe not entirely anatomically accurate—but it doesn't matter. What matters is that they are recognisable as hands, and the interaction modality is clear (in this case with a tangible interface).

On the rare occasion that you might want to spend a bit more time on hands, perhaps because you are using the sketch in a presentation, assignment, or research paper, you can make the basic sketch in pencil and then commit to pen and colour when you are fairly confident of the form (Fig. 5.11), although of course we prefer you just to sketch in pen directly, but hands are fairly tricksy.

Practical Application Tips

- Luckily, we all have hands, try sketching your nondominant hand in different poses and interactions.
- The Internet is your friend; if you are stuck or need to sketch a particular interaction, check visual references of people in natural settings.
- If you have one or two hand sketches you particularly like, scan them and reuse them!
- Sketching in HCI rarely will require photorealist hands; just the impression of a hand might be enough…

5.4 A Handy Guide

Fig. 5.10 Mixed media hands interacting with a tangible interface. Fineliner pen on paper then digitised using *Adobe Photoshop*. Miriam Sturdee, 2018

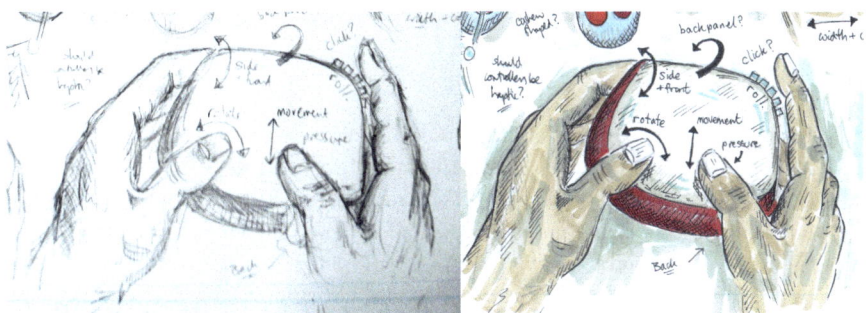

Fig. 5.11 Close up of sketch of novel game controller as part of an exercise in ideation. Left shows the initial pencil version, right depicts the "final" sketch. Pencil on paper and fineliner pen and marker on paper. Miriam Sturdee, 2016

5.5 Action Stations!

It is time to get your figures active! We've looked at the basic form for sketching people, but people rarely just exist standing alone on a blank page; people love to interact with things! However, this means noting how and where limbs bend, faces turn, and external factors influence them. In Fig. 5.12, Makayla has put in handy nodes to show where joints should be, and a handy one in the person's "core" to show the centre where somebody might pivot or bend their torso.

Fig. 5.12 Stick figure joints. *Procreate* App on *Apple iPad Pro* using *Apple Pencil*. Makayla Lewis, 2019

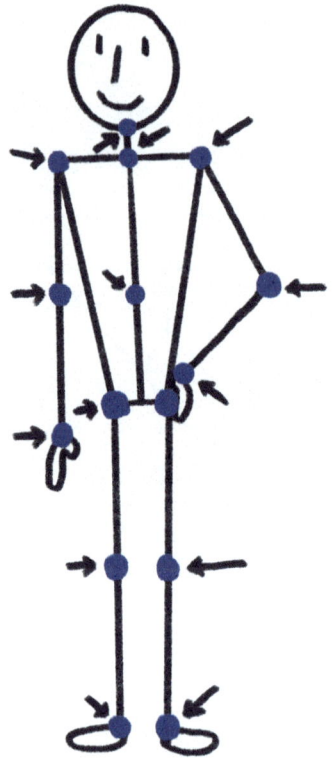

Representing movement is hard and requires practice for it to look natural (try Activity 5.5 more than once to start getting a feel for it). Luckily, when sketching simple figures such as the blob person, we can take some liberties with realism, as the intent is more important than precision, and with simple line arms and hands, we can approximate the interaction we wish.

To make the intent more obvious, we can also add annotations such as lines and arrows to show direction of movement or effort, such as the jumping person in Fig. 5.13 with the spiral underneath them, or the running person with the three straight lines indicating direction and speed. With stick figures especially, it is important to exaggerate the limbs at times to make sure the movement is clear. For example, if you look at the left middle person, the raised, curved arms indicate a huge jumping effort, even without movement lines to enhance the meaning.

In Fig. 5.14, Miriam has fleshed out the simple figures and utilised the nodes from Fig. 5.12 to figure out where the bends in limbs should be. Rather than add faces and clothes, the emphasis here is on the body; gaze direction is simply indicated with a crossed line; feet are triangular and hands largely ignored. Grey highlighting helps further indicate limb direction and stance; these extra details do not take long to add but help the viewer to appreciate the tension and movement of the fighting figures.

5.5 Action Stations!

Fig. 5.13 Examples of common stick and star person figure actions. *Procreate* App on *Apple iPad Pro* using *Apple Pencil*. Makayla Lewis, 2018

Practical Application Tips

- Try to practice angles and viewpoints to become confident with people looking and moving in all directions.
- Attending life-drawing classes can help you think about musculature and physiology, a great way of improving your overall sketching, too!
- In public, try 15-second drawings of people moving, using stick lines to capture the initial pose and then fleshing them out and indicating pivot points and nodes.

Now try the hands-on activities below!

5.6 Hands-On Activities

Activity 5.1: Sketch Your Neighbour (Pair Activity)
Learning objective—Engaging with abstract facial features and warming up
Time—Up to 5 minutes
Materials—Black fineliner, sketchbook or A4 paper, table surface or clipboard
Procedure:

- Find a person to work with (or use a mirror if you are alone).
- Place the nib of your pen around the centre of the paper.

Fig. 5.14 Two MMA experts perform a live demo, section of a live sketchnote. Fineliner pen on paper. Miriam Sturdee, 2019

- Look up, and gaze directly at the person you are sketching, and do not look at the paper.
- Now sketch the person's face, trying to add as many features and details as possible.
- When you both think you have finished and captured all of the person's facial features, hair and neck, stop sketching and look at your page.
- You have just created a beautiful, abstract facial sketch! Note where the features are and how they overlap or are separated—enjoy the freedom of sketching without looking.
- Present your portrait to your partner as a gift—or, as a class, put them all out on a table or stick to a wall to view as a group.

This is a popular activity for warming up at workshops and conferences; some versions utilise a paper bag to draw inside of to make sure nobody can look; others suggest you move around a room and try to capture as many people as possible,

Fig. 5.15 Sketch of Makayla. Fineliner pen on paper. Miriam Sturdee

and then the person being sketched can choose their favourite. Feel free to adapt and engage the activity as you feel works best! (Figs. 5.15 and 5.16)

Activity 5.2: Revisiting the Face Grid (Individual Activity)
Learning objective—Sketch a variety of expressions and hairstyles
Time—10 minutes
Materials—Black fineliner pen, sketchbook or A4 paper
Procedure:

- Quickly sketch nine ovals in 3 by 3 rows.
- Just above the middle of each oval, add two large dots—these are your eyes.
- Above each row and to the side of each row, add a straight line, bottom-facing curve, and a top-facing curve (see Fig. 5.17 for reference).
- Now, using the straight and curved lines, fill down and across. These are your eyebrows and mouths.
- Next, sketch a further three ovals with dots for eyes.
- Add a short straight line starting just below the eye dots in one oval and two similar lines at opposing 45-degree angles in the others. These are your noses!
- You now have a quick and easy facial expression reference grid.
- Next, refine the expressions over nine NEW ovals, this time separating eyebrows into two distinct features, adding more distinct noses, and trying open mouths, or even lipstick!
- Finally, add some funky hairstyles and headwear to complete your new grid.

Fig. 5.16 Sketch of Miriam. Fineliner pen on paper. Makayla Lewis

Fig. 5.17 Revisiting the face grid. *Procreate* App on *Apple iPad Pro* using *Apple Pencil*. Makayla Lewis

5.6 Hands-On Activities

Having completed this, you now have a useful face grid if you want to add expression to your sketches of people. Stick it to your wall or keep in a safe place to look back at. You will likely find that your skills with faces improve as you continue your sketching practice.

Activity 5.3: Emoji Faces (Individual Activity)
Learning objective—Experiment with unusual expressions, add features to basic faces
Time—15 minutes
Materials—Black fineliner pen, sketchbook or A4 paper, coloured pens or pencils as required
Procedure:

- Find an emoji face either on the web or on your phone that has a more unusual expression than the ones we have practiced so far (e.g. the thoughtful face, hard stare, brain exploding).
- Sketch the emoji as it is, paying attention to the quirks of eyes and mouth and any additional features such as hands or (in some cases) exploding brains.
- Pick another unusual emoji face, and repeat the procedure.
- When you have five emoji faces and using either your fineliner or colours, add different features to each, such as hair, ears and earrings, hats, sunglasses, collars, and where appropriate, upper body.
- Now you have five interesting characters! Finish the activity by adding a speech or thought bubble reflecting what they might be thinking.

This is a great way to experiment with faces when you don't have a live reference or are wanting to move away from self-portraits! You can also further extend the activity by adding full bodies, and reuse the expressions in the rest of the activities in this chapter.

Activity 5.4: Look at That! (Pair Activity)
Learning objective—Learn how to sketch gaze direction in faces
Time—10 minutes
Materials—Black fineliner pen, a pencil, a sketchbook or A4 paper. You may use Fig. 5.4 for reference
Procedure:

- Sketch nine basic ovals for faces in a row as you did for the face grid in Activity 5.2, but do not add features yet…
- Using Fig. 5.4 for reference, you will draw a curved cross in pencil or a light-coloured pen on each face.
- The centre face will have a straight cross, as this is the person looking straight toward you. For the other faces on this row, move the centre of the cross to the left or the right.
- For the rows above, change the centre of the cross to top left or right.
- For the rows below, change the centre of the cross to bottom left or right.
- The intersection of these crosses depicts the angle of the face (Fig. 5.18).

Fig. 5.18 Quick sketch of ovals with directional crosses. Fineliner pen on paper. Miriam Sturdee

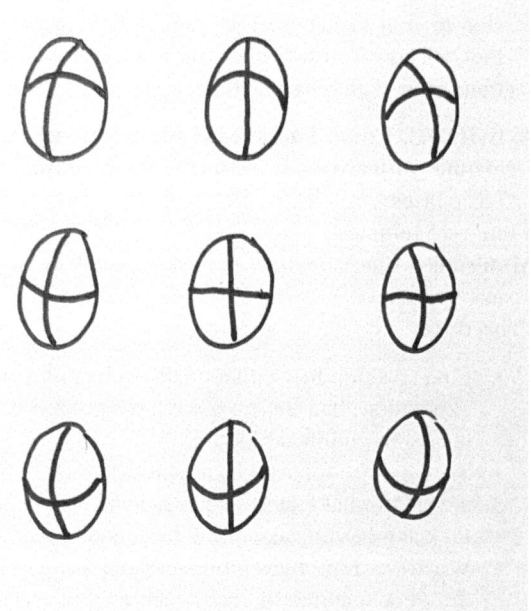

- Now sketch some eyes; place them at either side of the middle line, in the top half of each cross.
- Now add a mouth in the bottom half of the face, angled either left or right, up or down.
- Now add other facial features, starting at the top of the nose just underneath the cross, and then choose a mouth, and add some hair and/or ears!
- Repeat the activity for the BACK of the head.
- Now repeat the activity for more EXTREME viewpoints!

This is one of those skills that comes with practice. You can also approximate gaze by simply drawing a basic nose on an oval and scribbling in features. When you are sketching quickly, very sparse lines can indicate direction of gaze (Fig. 5.19). You can further this activity by adding objects for your people to look at and drawing a line between the object and eyes.

Activity 5.5: Interact! (Individual Activity)
Learning objective—Learn how to depict interaction with objects and devices
Time—15 minutes, 5 minutes per object or device
Materials—Black fineliner, light-coloured pen or pencil, sketchbook or A4 paper
Procedure:

- Think of three objects or devices that you use regularly and enjoy using (e.g. mobile phone, stylus, tablet; mug, fork, dial).
- Lay them out in front of you for reference, or sketch from memory.

5.6 Hands-On Activities

Fig. 5.19 140 doodle people in action. *Procreate* App on *Apple iPad Pro* using *Apple Pencil*. Makayla Lewis, 2019

- Quickly sketch the object as a basic icon (see Chap. 3) using the coloured pen or pencil.
- Now pick up the object and notice which parts of your hand are visible and what angle the fingers are. Put the object down, and sketch the rough shape of your hand holding the object using the black fineliner.
- Is the object or device a two-handed interaction? Use your dominant hand to hold or point, depending on the style of interaction; keep the position in your mind, and quickly add the rough shape of the second hand, again using the fineliner.
- Repeat the previous two steps with your other two objects.
- Now you have three objects and three rough "realistic" hand sketches.
- Referencing back to Fig. 5.8, try sketching four of the different hand styles interacting with the same object.
- You should finish up with five hands interacting with each object.

Sometimes you will find that a simple shape or finger and blob for a hand is enough, or for more complex interactions or close ups, you may need to have the correct number of fingers and thumbs. We'll look more at how to sketch for different levels of close up in Chap. 6.

Activity 5.6: All Together Now (Group Activity)
Learning objective—How to sketch people doing-saying-moving, the ability to sketch a selection of heads and bodies and hands
Time—30 minutes
Materials—Black fineliner pen, thicker black marker pen, A4 or A3 paper, scissors
Procedure:

- Form groups of four, two people will be posing, and two people will be sketching.
- The pair who are posing should choose an interaction (laughing together, holding hands, sitting and standing—the choice is open!)
- The sketching pair should both sketch the other group members in a style of their choosing, as large as possible, on a piece of A3 paper, without drawing the facial expression.
- The pairs should now swap roles.
- Together, examine the images; what worked well? What style of face and hands were used?
- All group members should now collaboratively sketch out a face grid on another sheet of paper, giving each person different expression, hair, and accessories, and then cut each face out using the scissors. Make sure the faces are of a similar size to the previous drawings.
- Now place the new expressions one by one over the faces of the interactive pair sketches. How does the expression change the meaning of the pose?
- Repeat the activity, but using hands instead, of a similar fidelity to the figures.
- Each member of the group should choose their favourite combination and add speech or thought bubbles, representing what they think is happening in the scene.
- Finally, each sketch should have context added, objects, background, and props.

This is a form of paper prototyping for people; you can redesign the interactions as often as you like. If you want to take the activity further, you can separate out parts of the body and play mix and match with all sorts of different ideas. You could also use this to build a storyboard (see Chap. 6), by either using the poses as a sketching reference for each frame or photographing each one and building up digitally (though of course we advocate sketching!).

References

Books, Papers, and Articles

Lengyel, D., Kaipainen, K., Sturdee, M., Heron, M., Lewis, M., & Liddle, J. (2023). *Hands-on workshop on tabletop role-playing for inclusive design: Imagining sustainable futures for 'older adults'*. Academic Mindtrek.

Mondloch, C. J., Lewis, T. L., Budreau, D. R., Maurer, D., Dannemiller, J. L., Stephens, B. R., & Kleiner-Gathercoal, K. A. (1999). Face perception during early infancy. *Psychological Science, 10*(5), 419–422.

Websites

W1 Randall Munroe (formerly of NASA) is an expert at making simple stick people without faces emotive and interesting! www.XKCD.com
W2 Useful reference colours for skintones – www.skintone.google/getstarted

Further Reading

Campanario, G. (2014). *The urban sketching handbook: People and motion: Tips and techniques for drawing on location.* Quarry Books.
Charles Stephen. (2008). *Draw figures in action (Draw books).* A&C Black Visual Arts. Revised edition.
Fox, T. (2022a). *Fundamentals of character design: How to create engaging characters for illustration, animation & visual development.* 3dtotal Publishing.
Fox, T. (2022b). *Anatomy for artists: Drawing form & pose: The ultimate guide to drawing anatomy in perspective and pose.* 3dtotal Publishing.
Hart, C. (2009). *Figure it out!: The beginner's guide to drawing people.* Sixth & Spring Books.
Hart, C. (2016). *Figure it out! Drawing essential poses: The beginner's guide to the natural-looking figure.* Sixth & Spring Books.
Hart, C. (2021). *Figure it out! Faces & expressions: The ultimate drawing guide for the beginning artist.* Sixth & Spring Books.
Umoto, S. (2010). *Illustration school: Let's draw happy people.* Quarry Books.

Chapter 6
Exploring Visual Narratives

6.1 What Do We Mean by "Visual Narrative"?

There are many ways to tell a story with pictures—or, in this case, sketches! You'll hear this described in many ways: storyboard, comic, vignette, illustrated scenario, sketchnote, and so forth. In this chapter, we describe the basic process of telling a story with images as a "visual narrative" because it can encompass all of the above and allows us to share the process without getting *too* bogged down in terminology. There are some distinctions which might be helpful to share however, so we will describe them in this chapter.

A visual narrative is simply a way of visualising a story, a sequence of events. Comics can also be described as scenarios—what we call "visual narrative" differs between practitioners and fields. Someone working in UX will likely have a storyboard or scenario; an illustrator or artist might make a comic (or a graphic novel if it is very long!). In reality though, there are no hard distinctions in terms of content, as long as the meaning and story you are trying to convey is clear and you frame your work appropriately for the domain it will be used in.

Telling stories is a distinctly human activity. Prior to reading and writing being taught, people told stories using the oral tradition, acted, or drew images to show a sequence of events—the oldest of these are thousands of years old and adorn the walls of the caves of our ancestors. Even modern films are usually sketched out as a sequence of static images at first, to help the director and actors decide what will happen and how the shots will be composed.

Mastering visual storytelling is a valuable skill. It can help people understand interactions, use cases, and the wider context of a technology or application. It can help you communicate complex concepts to your peers, stakeholders, and clients. You can even use visual narrative to tell the story of research and publish new inquiry in "pictorial" form (see Chap. 11).

© The Author(s), under exclusive license to Springer Nature Switzerland AG 2024
M. Lewis, M. Sturdee, *Sketching in Human-Computer Interaction*,
https://doi.org/10.1007/978-3-031-50136-4_6

By the end of this chapter, you should be able to:

1. Construct an appropriate style of visual narrative using figures, objects, and places.
2. Apply an appropriate story arc to your visual narrative.
3. Learn how to construct scenes and frames and use different angles and zoom levels.

6.2 Vignettes

A vignette is a French word rooted in *vigne*—a vine. It used to (as in seventeenth century!) mean an image that was used in the border of a page in a book, or signalling the beginning or end of a chapter. The current definition of this in the Oxford English Dictionary (W1) suggests either "a brief evocative description, account, or episode" or "a small illustration or portrait photograph which fades into its background without a definite border". In this context, we prescribe to the former definition, as the brevity refers to a single image showing many things. Figure 6.1 gives an example of a vignette where multiple interactions and storytelling elements are contained in a single frame—in this case with minimal text.

A vignette in HCI might be used in a conference poster, a report, or a research paper (where space is often at a premium) or could also refer to a "visual abstract" (see Chap. 11). See Fig. 6.2 for another example where the author has used an annotated vignette to describe the place they live. This example could also be described as a "sketchnote" of a place. Figure 6.3, in contrast, is a very loose sketched depiction of thoughts and demonstrates that multiple approaches can be used to single-page or single-image storytelling. The first two examples have been planned to some extent, and composed, whereas the organic nature of the third invites interrogation; the first two appear as "complete".

Try both methods for yourself and decide in what situation you might use either:

- The planned and "finished" version.
- The loose and organic version. How does your own personal style best fit?
- Which do you prefer?

6.3 Sketchnotes

6.3.1 What?

When we take notes, we save interesting points and ideas from a talk, lecture panel, workshop, experiment, or participant to return to later. Low-fidelity sketches often help to express experiences and complex content. Sketchnoting is a technique where

6.3 Sketchnotes

Fig. 6.1 Vignette—a day in the life… *Procreate* App on *Apple iPad Pro* using *Apple Pencil*. Makayla Lewis, 2023

Fig. 6.2 Vignette—A visual and annotated description of living in Lancaster and its surrounds. Fineliner pen on paper. Miriam Sturdee, 2019

6.3 Sketchnotes

Fig. 6.3 Vignette—AI presentation at Kingston University. *Procreate* App on *Apple iPad Pro* using *Apple Pencil*, Makayla Lewis, 2023

you take notes in a visual format that incorporates both sketches and annotations. Sketchnotes can enhance these sketches by including simple connectors, containers, and separators with consideration of structure and style. When we craft sketchnotes, we add sketched visual elements to those points and ideas, whether as simple as emphasising text or adding icons and thematic references to the recorded item.

6.3.2 Why?

The critical part is the ideas and thoughts you capture and develop your best style. Sketchnotes offer you an opportunity to:

- Improve your retention of information.
- Increase your creativity and sketching skills in various contexts and environments.
- Gain better focus and engagement in meetings, lectures, workshops, conferences, and presentations.
- Help you to develop a deeper understanding of complex concepts, ideas, and experiences.
- Improve your critical thinking skills in various contexts and environments.
- Improve your networking with the presenter and audience.

Fig. 6.4 Sketchnotes from CUX meetup "The ethics of voice 1st interfaces". Fineliner pen and marker on paper. Makayla Lewis, 2022

6.3.3 Where?

In HCI, sketchnotes are often used in a variety of environments and for different needs; these include but are limited to:

Teaching and Learning Sketches are a powerful tool for enhancing student learning and engagement in the classroom by allowing lecturers and students to create visual representations of information; it can help students to develop and understand a deeper understanding of complex HCI concepts and ideas, e.g. Figs. 6.4 and 6.5. Furthermore, creative HCI modules can also be used as a form of alternative visual assessment to allow students to demonstrate their knowledge and critical thinking through coursework.

Research Findings HCI research often includes telling a story about the user's experience and feelings when using a system. Sketchnotes can support researchers and users to cocreate an easy-to-follow visual story for stakeholders that summarises research findings in a more user-centric, relatable, and memorable way, e.g. Makayla's 2014 research on meta-stories (see W2).

Lo-fi Prototyping Sketchnotes can support the creation of quick low-fidelity prototypes (paper or digital design concepts) to communicate their design decisions to colleagues or gather user feedback to improve design decisions, e.g. Figs. 6.6 and 6.7.

Fig. 6.5 Cyber risk sketchnote—the digital butterfly effect with Craig Templeton. Fineliner pen and marker on paper. Miriam Sturdee, 2016

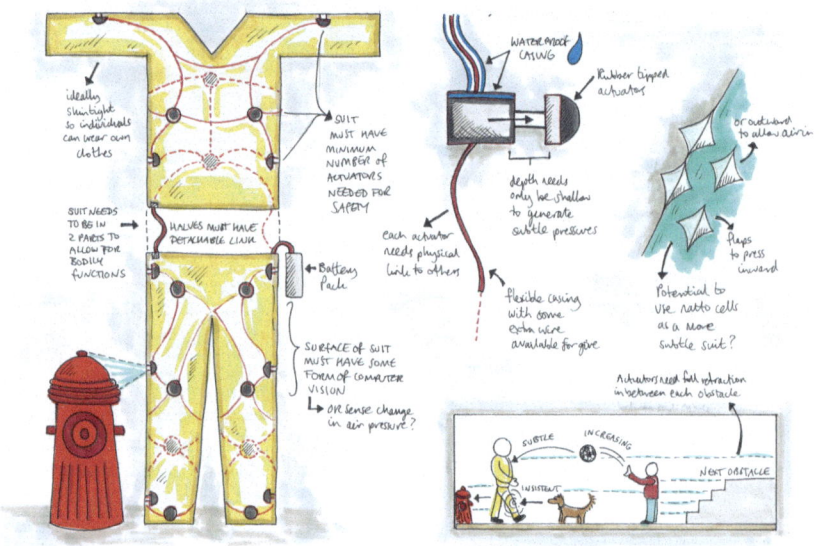

Fig. 6.6 A distance-to-touch-suit idea to support blind people in navigating everyday spaces, using cloth with embedded actuators. Fineliner pen and marker on paper. Miriam Sturdee 2018

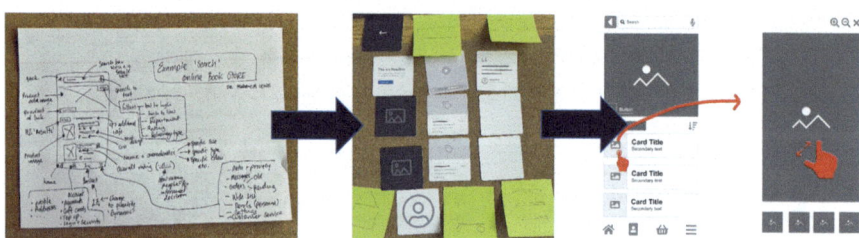

Fig. 6.7 Live demonstration of the role of sketching in the low-fidelity to mid-fidelity process as part of a Kingston University London MSc User Experience module – Design Thinking Theory and Practice. On-the-Go Photograph, fineliner pen on paper; fineliner pen on Post-it Notes; and *Miro* Whiteboard user experience wireframe elements. Makayla Lewis, 2023

User Feedback Sketchnotes are often used to capture user feedback and insights during problem identification, user research, ideation, and testing because there are always users who may struggle with text or verbal utterances to share their experiences, pain points, and/or suggestions, e.g. Fig. 6.8.

Focus Groups and Workshops Sketchnotes can enhance data gathered from user research by making discussions more engaging and memorable for participants. Miriam and Makayla often use such outputs for stakeholder reports and lay designs for participants to make the data easier and quicker for the reader to understand, e.g. Figs. 6.9 and 6.10.

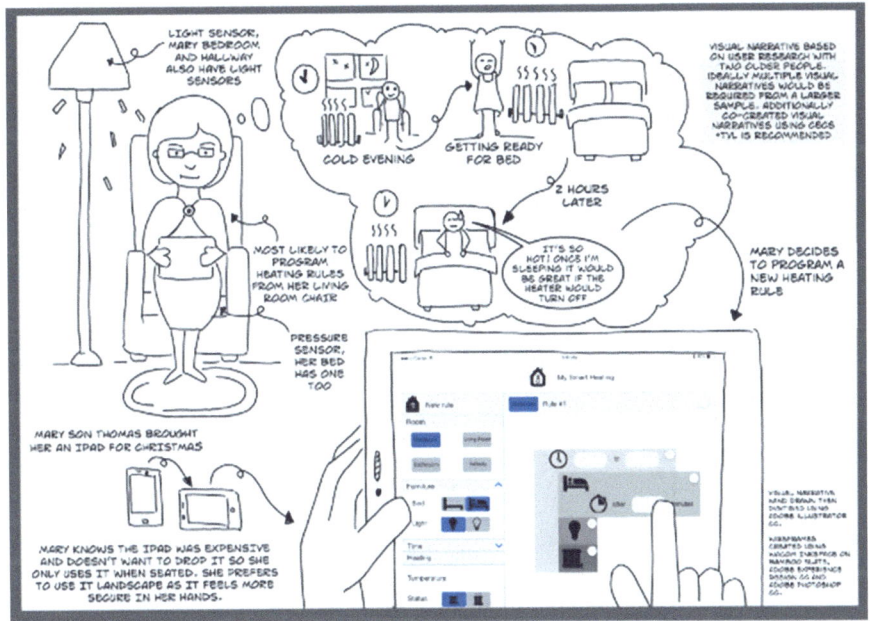

Fig. 6.8 Visual representation of a smart heating IoT and App in a house for an older person. *Photoshop* on *Microsoft Surface Pro* using *Microsoft Surface Pen*. Makayla Lewis, 2018

Fig. 6.9 Makayla's approach to sketchnoting a live TTRPG workshop looking at sustainability in older adults. Fineliner pen and marker on paper (Lengyel et al., 2023)

Fig. 6.10 Miriam's approach to sketchnoting a live TTRPG workshop looking at sustainability in older adults. Fineliner pen and marker on paper (Lengyel et al., 2023)

Dissemination Sketchnotes can enhance the readability and accessibility of HCI research to users and stakeholders who may struggle with (or not prefer) text-based reports, e.g. Fig. 6.11.

Collaboration Sketchnotes can support collaborative research, especially amongst interdisciplinary and/or varied first-language teams; the outputs can ensure that all concerned have a shared understanding of the project goal and findings.

Sketchnotes can also be used to describe a process, story, or even a visit to a museum or art gallery! You can use them to describe your thoughts like an elaborate mind-map, make reminders to help revise for an exam, or even organise the ideas of a group or team as you brainstorm and project.

There is a huge sketchnote community in the world, with books, events, and meetups. We've put some useful resources in the references section at the end of this chapter. Regardless of how you choose to use it, it is important to ensure they are accessible (see Chap. 9).

Fig. 6.11 A digital sketchnote for the Design Council. *Procreate* App on *Apple iPad Pro* using *Apple Pencil*. Makayla Lewis, 2021

6.3.4 How?

Sketchnotes are usually also one-page visual narratives, but they do not conform to the same conventions of scene construction as a vignette. In fact, most sketchnote artists have completely different approaches and are constantly reinventing or adapting their style—see Figs. 6.11 and 6.12 for an example by Makayla and one by Miriam; how do they differ? How do they differ from our other sketchnote examples in this chapter?

Many sketchnotes make use of icons and connectors to illustrate concepts and show the flow of the narrative—this is where all of your knowledge from Chap. 3

Fig. 6.12 Sketchnote about a start-up "Repairly" utilising the circular economy. Fineliner pen and marker on paper. Miriam Sturdee, 2017

will come in useful! Remember when we told you to practise domain-specific icons you would use regularly? Sketchnotes are the perfect place to use them.

You can use a sketchnote in almost any context. The most common form is when the sketcher is listening to a talk or lecture by someone. These sketchnotes usually have a small "portrait" of the speaker somewhere on the page—if you're worried about drawing the speaker, find out who it will be and sketch them from a photograph before the event. You can also plan your page and style or theme before you start which will help keep you on track so you can focus on the main content (see Fig. 6.13 for example approaches to your page). Your practice sketching people in Chap. 5 will help with the portrait—if in doubt, keep it simple, and flatter the person.

Some common elements of a sketchnote are a title, portrait, connectors and separators, text in different forms/weights, icons, borders, speech bubbles, thought bubbles, small diagrams…; you will figure out what you need from both the context and during the sketching flow. There is no right way to do this! Sketchnoting is a perfect way to refine your style and express yourself.

We advocate pen and paper sketching, but digital sketchnotes are very common, especially within the HCI community. One great thing about the digital approach is the ability to layer up lighter colours on a darker background or write in white over black or a colour. Of course, you can also "undo" things you don't like—but you will probably find that you lack the time during full sketching flow! Digital sketchnotes also can be easily shared in high resolution (Fig. 6.11), which is great for fast documentary of an event, without the difficulties of photographing your page in poor or unreliable lighting conditions (see Chap. 14 for advice on photographing and scanning your sketches).

Practical Application Tips for Sketchnotes (Individual)

BEFORE the Event
- **Choose appropriate tools**—Your favourite sketchbook, hardback is recommended as you may need to sketchnote on your lap (there may not be tables) (see W2). A good sketchnote set contains an easy-flow black fineliner or slim pen, a mid-grey highlighter marker (which doesn't bleed through paper too much), and another colour for emphasis.
- **Practice icons**—Remember Chap. 3? Well, this is where your icons will come into their own. Practice those that you think will come up a lot, for example draw icons representing IOT devices if the event is about that domain. This will help you during your sketchnote as you can quickly call upon prior knowledge. Oh, and don't get us started on ink… Makayla has started a sketchnote so many times and run out of ink, thus, she always brings a refillable fineliner, e.g. a *Copic Multiliner SP*, and the refill for it, so it can be filled on-site if needed.
- **Colours**—Use black fineliner pens for outlines (line art) and text, one pastel or light shade marker to highlight important points, and a grey colour marker for shading.
- **If you are using digital**—A fun trick is to block out some colour, grey or black, and write over it in white text—visually, this looks great, and helps you to ensure the size of the elements is consistent; it also helps you know where you are when you are sketching on a zoomed-in digital canvas!

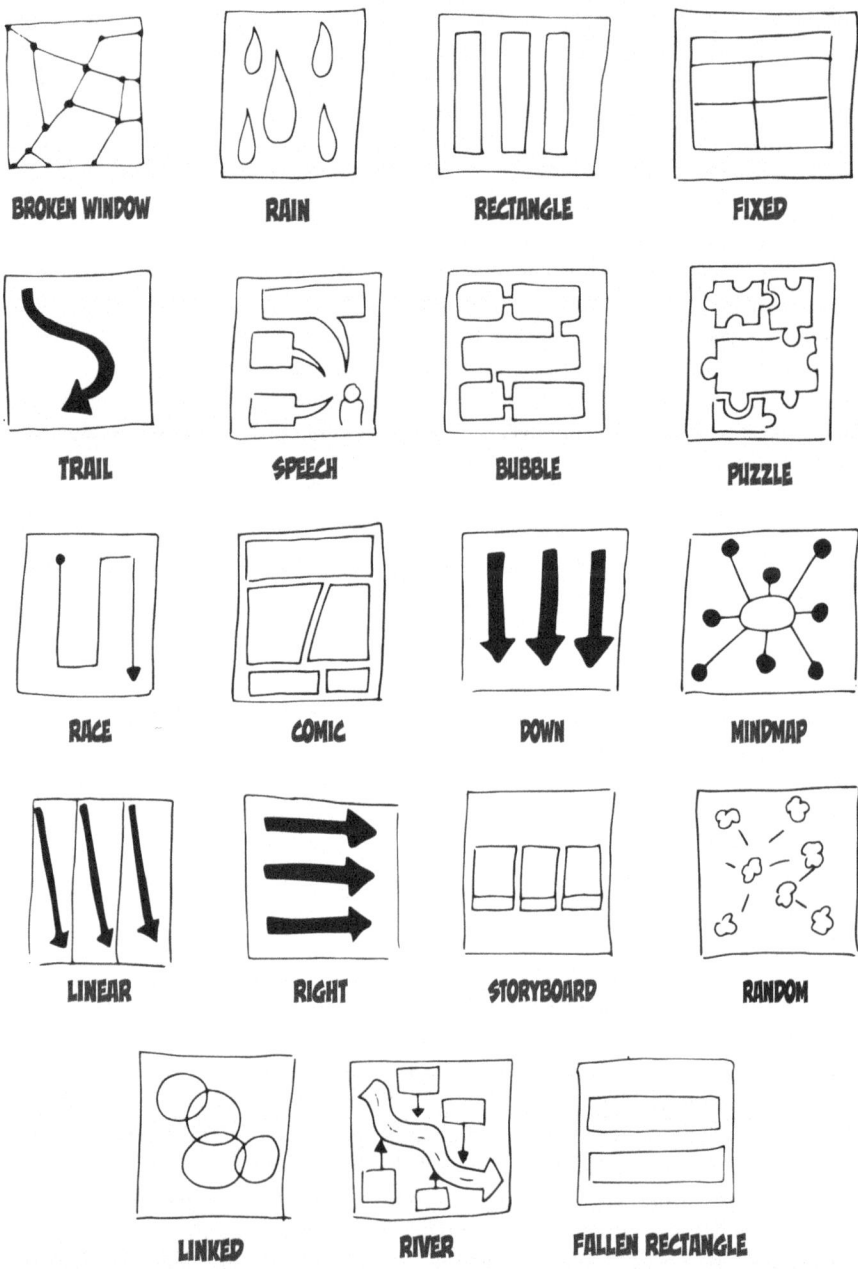

Fig. 6.13 19 Sketchnote styles. *Photoshop* on *Microsoft Surface Pro* using *Microsoft Surface Pen*. Makayla Lewis, 2016

DURING the Event
- **Arrive early**—If you are sketchnoting on location (rather than attending a digital event or working from a static environment or resource), arrive early and get a seat that gives you a good view of the action (people, slides, or artefacts).
- **Downtime**—Use downtime at the beginning of an event to organise your page and decide on an approach; this is also a good time to sketch a portrait of the person giving the talk and the title.
- **Content**—Don't try to capture EVERYTHING on paper; sketchnotes are subjective accounts of an event or topic; choose what interests YOU, it will ensure you do not become overwhelmed.
- **Digital sketching**—Digital can be cool for sketchnotes, but beware of the temptation to erase and start again! Sketchnoting waits for no one.
- **Verification**—Use question and answer time to verify information, i.e. to fill in the gaps, (incomplete areas), or if you are unsure about something, ask the speaker a question.

AFTER the Event
- **Colour**—We recommend only colouring your sketchnote after the event (when you finish your line art). Try to finish your sketchnote in one sitting—it is hard to return to sketchnotes after an event; the inspiration and adrenaline wear off (Fig. 6.12 was coloured later).
- **Share with the speaker(s) and audience**—Use your sketchnote as a boundary object to support discussions and networking.
- **Archive and Revisit**—it's a visual record (ledger) that you can return to; don't forget about it.
- **Keep practising**—You will quickly progress and fall into a personal style that becomes your "go-to"—as much as you find your people, objectives and lettering will start looking more consistent, so will your sketchnotes.

For discussion, examples and tips for crafting collaborative sketchnotes with others, please see Chap. 12. Try sketchnoting next time you are in a lecture, or meeting. If it is online or remote, you can also make sure you have a comfortable set up for your practice. The websites (W3–W6) in the references section showcase some wonderful sketchnotes by leading practitioners.

Also try sketchnoting for yourself by completing Hands-On Activity 6.2.

6.4 Comics

There might appear to be a lot of overlap between a comic and a storyboard, but, as you will see in the next section, there are big differences! A storyboard could map out a story that will later be actualised in another way, perhaps as a film or animation, or will communicate a concept, user scenario, or user journey map. As a storyboard creator you will often follow a set of commonly used guidelines and practices, and if you are a comics reader, you will have expectations of consistency, especially in terms of reading order e.g. Manga: a reversed Z and English comics, left to right, however, in terms of styling, there are no rules when it comes to making comics! Comics are as much an expression of personal style, imagination, and technique as a work of art and are often celebrated as such. Anything goes, from humorous web comics drawn digitally, to serious sketched black and white pen imagery in a paper zine. Comics can have branching narratives, be one panel, a page, or span hundreds of pages in a graphic novel.

The history of comics is rich and deep and as such is beyond the scope of this book, but we recommend reading "Understanding Comics" by Scott McCloud (McCloud & Martin, 1993), "Comics and Sequential Art" which celebrates the works of Will Eisner (Eisner, 2008), and also "How Comics Work" by Dave Gibbons (Gibbons & Pilcher, 2017). Figure 6.14 also showcases some of the background of comics and graphic novels in sketchnote form (note that the style used is much more "sketchy" and also represents choice of style specific to the topic). We invite you to explore comics in your own way, and using your own sketching style!

Within HCI, comics are becoming more popular and have been used as a focus of inquiry, a method, as outputs that help the reader engage visually with research materials, and all of the above. We'll talk more about research-based sketching in HCI in Chap. 11. As some examples of comics in HCI, we have included some of our own here. The first two images (Figs. 6.15 and 6.16) are part of a group research

Fig. 6.14 Introduction to Comic Art and Graphic Novels, a sketchnote of Benoit Peeters' inaugural professorial talk at Lancaster University. Fineliner pen and marker on paper. Miriam Sturdee, 2016

6.4 Comics

Fig. 6.15 Without words… full page comic introduction with "alt narrative". Fineliner pen and marker on paper. Miriam Sturdee 2021 (in Lewis et al., 2022)

exploration of comics without words (Lewis et al., 2022), specifically designed to showcase how we might elaborate on "alt text" (see Chap. 8 on Accessibility for Sketches for more information) to create "alt narratives". These narratives are designed to give those who use screen readers as rich an experience as those who engage visually with content. In particular, the first example comic (Fig. 6.15) represents the introduction to the research paper, whereas the second (Fig. 6.16) depicts a research conference event on sketching that is documented in comics form.

Fig. 6.16 Extract from HCI comic summarising a workshop at ACM SIGCHI 2019. *Procreate* App on *Apple iPad Pro* using *Apple Pencil*. Makayla Lewis, 2021 (in Lewis et al., 2022)

In contrast to Figs. 6.15 and 6.16, which are both research paper and comic at the same time, Fig. 6.17 is a sketched-comic created as part of an exercise in "world building" which is a concept from design fiction (see Chap. 7). It situates its story around a piece of fictional technology in use—the empathy detector—and how this might cause trouble in the future of Internet dating (Sturdee et al., 2016). This comic stands alone as a visual narrative but is also part of a bigger research picture. However they are used, comics can be an important part of the research narrative, and whilst some of these might look "polished", they all started life as sketches.

6.4 Comics

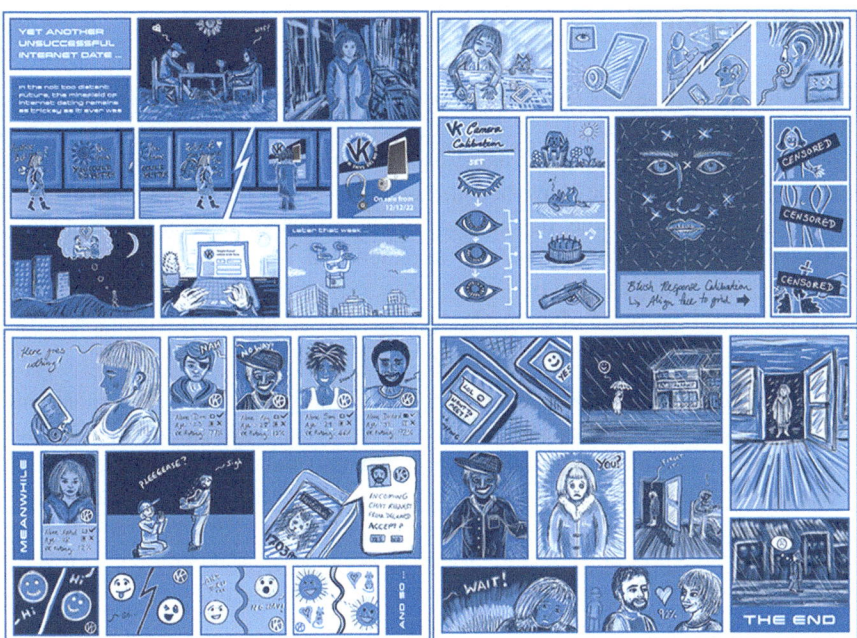

Fig. 6.17 A four-page comic depicting a love story set in a time of "digital empathy" (Sturdee et al., 2016). *Adobe Photoshop* on *Cintiq Companion 2*, Miriam Sturdee, 2016

Practical Application Tips

- Don't try to force yourself into the style of your favourite comic artist; it will create unnecessary stress and comparison. Instead, try to sketch out the story in your own developing style.
- If you find a scene difficult to sketch, approach the comic in stages, and pull separate images together digitally afterwards, or even create a paper-based mixed-media piece (you could use Post-it notes)!
- Creating a likeness of someone is very hard; in a comic, you may find yourself drawing the same character several times over—start with simple faces and bodies as you build confidence—use tracing paper or digital layers to ensure consistency.
- Think outside of the box (or frame)—try designing a page with odd-shaped frames or even none!
- Leave space for text—it is easier to write the script and dialogue in advance so you know how much of the frame the speech bubble or caption will take up (see Fig. 6.20 in the next section for an example of planning your narrative).
- Have fun! Comics are a pure form of storytelling for enjoyment and ease of communication.

Try making short comics yourself by completing Hands-On Activity 6.2.

6.5 Storyboards

Similar to a comic, a storyboard is a visual representation of a linear sequence arranged together to visualise a story that is fiction, nonfiction, or a mixture. Popularised in the 1920s by Walt Disney Studios (Disney Editions, 2008 and 2011a, b; and Williams, 2012), they have become an essential tool for many creative fields, e.g. film, animation, game design, and broader design specialties, including HCI. Storyboards are informal and easy to understand, reiterating the mantra "only draw what you need" (Chap. 1).

In HCI, storyboards are often used to plan, visualise, and communicate a narrative or series of actions in a user-centred way:

- **Journey**—Story of a user or community that visually depicts how a user or community interacts with an application, software, website, system, or intervention, with the aim of identifying challenges (pain points) and optimising their journey.
- **Scenarios**—Explore different scenarios or use cases for an application, software, website, system, or intervention.
- **Evaluation**—An opportunity for users, colleagues, or stakeholders to review and provide feedback on the current or future state of a current or future application, software, website, system, or intervention.
- **Documentation and dissemination**—Support understanding of (or to share) current and/or future application, software, website, or system by guiding the reader through a story (e.g. Figs. 6.18 and 6.19).

Fig. 6.18 Daily TodaysDoodle storyboard by Makayla, a good example of regular practice for scene creation! Fineliner pen and marker on paper. Makayla Lewis, 2012

6.5 Storyboards

Fig. 6.19 Example rough storyboard (left), and the final version during inking (right). This was later developed into a full-page comic for ACM Interactions magazine's new sketch-based feature (Sturdee, 2023). Note the visible differences between the first draft and the final, which was signed off by the editors. Fineliner pen and marker on paper. Miriam Sturdee, 2023

Storyboards are a stepped story with a start (introduce the scene), middle (event or interactions of note), and end (character has achieved goals). They will allow you to break down complex stories into short and comprehensible frames. The critical elements of a storyboard include the following:

Goal The first step in creating a storyboard is identifying its goal. To help you with this, we recommend answering the following questions:

- What is the purpose of your storyboard? (Why is it essential to visually represent an experience, flow/journey, process, or event?)
- Who will be the viewer (this will determine the language level of the storyboard, e.g. children, adults, etc.)?
- Do you have enough user and environmental data for the storyboard to be accurate and relatable?
- What will you do with the storyboard when it's finished? (Will it be digital or printed? What size will it be shared? Will it be for internal or external use?)

Plot Create a blueprint or rough script for your storyboard, it will ensure your story is clear and relatable. It should identify key moments/events, often referred to as "plot points"; it should have a beginning (what caused the story to commence), middle (build up to a critical moment), critical plot point (the key event, positive or

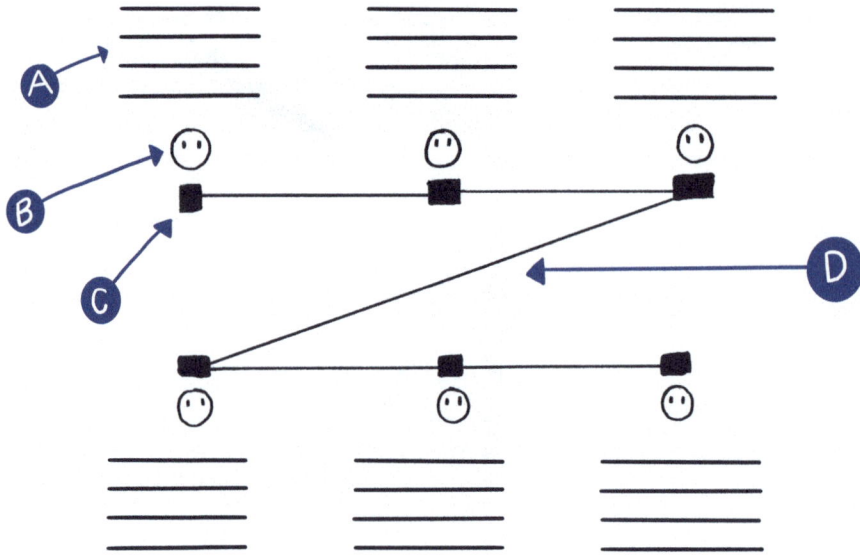

Fig. 6.20 A visual guide for creating a plot for your storyboard quickly and easily. Image key: (**a**) brief description of moment/event, (**b**) if there are any characters what they feel about the moment or event, (**c**) location of the event, and (**d**) the read order of the story. Feel free to adjust for a vignette, frame, and the length of your storyboard. *Procreate* App on *Apple iPad Pro* using *Apple Pencil*. Makayla Lewis

negative, that defines the story—the story highlight), and ending (how the story ends, open endings are not encouraged) (Eisner, 2008). The plot should identify the people, environment, actions, interactions, and emotions to ensure the storyboard is straightforward and relatable, e.g. see Fig. 6.20. It is recommended that the plot draft is reviewed by a colleague, user, and community before you progress to ensure its accuracy.

Characters Next you need to describe the people in your story, often referred to as characters. Think of your characters as simplistic personas (e.g. proto personas). In HCI, we often create storyboards that visually represent a group of people rather than one person; thus, consider your characters to be an archetype of the people you are telling a story about (your data). To help, you can ask yourself: Who is your main actor? What is their focus / purpose? Your rough character sketch should include their general appearance (body type, clothes, facial expressions, and favourite objects), age, and gender. However, we recommend your character gender is ambiguous unless the storyboard goal is directed at a particular community.

Format Deciding on the format of the storyboard and how plot points will be presented is our next step. A decision should be made early and kept the same after starting this phase, as consistency is essential. Will it be a vignette (one large image where the plot is intermingled, e.g. see Fig. 6.1) or frames (up to 8 boarded pictures separated by 1cm that are connected with or without arrows, e.g. Fig. 6.13)? Remember, sketches are plentiful, so if you dislike them, you can change to a different format. Remember not to change the style mid-story as it will confuse your viewer.

Building a Scene's "Shots" Your storyboard will contain many elements (characters and objects), all of which you should now be becoming familiar with—and hopefully, you are practising regularly! Placing characters and objects in a scene alongside realistic perspectives and a "camera" angle is critical to ensuring accurate and relatable scenes. You want the viewers to focus on the story, not a questionable composition, e.g. buildings are not the size of people unless they are made out of LEGO, smartphones are as big as the stick person's arms, heads are unnaturally large, etc. It seems minor, but the viewer will only notice these quirks and become fixated, thus forgetting about the story.

Time and Transition Indications of the time (duration) in plot points and between plot points will help readability and continuity. You must consider whether the each scene is in the past, current, or future. You could include the following:

- Changing objects, e.g. changing a clock time or the movement of the sun outside a window.
- Change in lighting, e.g. indoors or outside (drawn, day, dusk, or night).
- Change in weather e.g. sunny, cloudy, rainy, rainbow.
- Slowly adjust the character's location/positioning and/or object, e.g. a character holding a smartphone, placing the phone on a counter, getting distracted by something, and walking away; a character panicking about their smartphone location, then returning the "camera angle" to the smartphone alone.
- Dialogue that implies time, e.g. characters discussion about the past in the present with what could happen in the future.

Communication If your storyboard is about verbal discussions or interactions with an object, use speech bubbles; however, use them with caution. We recommend keeping the communication brief; only so much can be included in a plot point (vignette or frame). If the dialogue is extended, annotations or bullet points underneath the vignette or frame are recommended.

Colour and Highlights To draw attention to interactions or things of note either using placement or highlights, do not rely on black and white, e.g. if a character is holding a smartphone, which is where you want the reader to focus, colour the smartphone in a pastel or highlight shade, and leave the rest of the sketch uncoloured. You could also use pen over pencil to make lines bolder, thus creating a greater focus. Furthermore, focal characters, environments, or objects should be more detailed. Whilst non-focal characters, environments, or objects should be blurred, don't forget perspective, as with compostion you do not want your viewer to become fixated with ceilings that are too low unless your storyboard is about mystical giants.

Annotations Descriptive text (often called annotation) is recommended to add extra content to a plot point without making the sketch too busy to understand. We recommend using callouts for vignettes, and for storyboards formatted in frames, they should appear underneath. The purpose of annotation is to provide a supportive explanation of what is happening if the sketch alone does not convey the plot point, for example, character feeling, plot point purpose, additional dialogue, or if you have yet to sketch something well (do not worry, Makayla cannot draw hands well, and Miriam struggles to sketch buildings and vehicles). To help the reader in such situations, explain briefly so they are not confused; to help ensure you do not over-describe (long annotations), remember a popular British expression to people who talk in circles "Excuse me—make a long story short"—we recommend using concise bullet points.

Storyboards can be updated if their goal is adjusted as a research project progresses; they allow the creator to iterate and experiment; they are not static "one time only" artefacts. They are plentiful and easily changed.

Practical Application Tips

- Make sure your sketch output matches the intended recipient—rough visual stories invite interrogation! Polished visual stories invite reflection.
- Scenes are a series of icons—consider your icons before putting pen to paper.
- Use a variety of shots and camera angles—e.g. long shots from a distance, over the shoulder, close-ups…
- It's all about focus; what do you want the viewer to notice? Highlight them!
- Remember—again—ideas not art; you need to be able to convey the story to your colleague, user, or stakeholder (be it research or participant) keep it simple, only sketch what you need.
- In case you need to, photo tracing is an acceptable way of pulling in imagery—feel free to adapt and annotate (and attribute) or take your own photographs! (See Greenberg et al., 2012).
- Most importantly, keep practising; it does not need to be with data. It can be everyday life (e.g. Fig. 6.18) or fun prompts, e.g. the everyday life of Miriam's cat Ren as he re-shares his lovely hairballs and food with his beloved owner (Fig. 6.21).

6.6 Hands-On Activities

Activity 6.1: Build a Story (Individual/Pair/Group Activity)
Learning objective—To construct a vignette storyboard
Time—15 minutes
Materials—A sheet of A4 paper, plus basic materials needed, and any extra items that may be required (e.g. sticky notes, rulers)

6.6 Hands-On Activities

Fig. 6.21 Fun prompt storyboard, drawing inspiration from living with cats. The cat is unimpressed. Fineliner pen and marker on paper. Miriam Sturdee, 2023

Procedure:

- Recreate Figs. 6.22 and 6.23 on one piece of paper; feel free to adjust to your preferences. We recommend using a pencil.
- Using a pen to add to the background scene creates a funny and exciting story.
    (a) Remember, it is a vignette, not a framed storyboard; you must complete the story on 1 page, using the background as a guide.
    (b) Return to Sect. 6.6 to identify and define the plot and characters before you begin; it will make the process quicker, easier, and most importantly, clear and relatable for the person reading your story.
    (c) Once you have completed your scene, go over the background with a pen, and sketch around anything you have added to the scene.
    (d) Erase the pencil and colour the scene.
    (e) Give the vignette storyboard to a friend, colleague, or family member, and ask them to tell you the story.
- Their response will tell you how successful you were in telling the story! Feel free to repeat to improve if necessary.

Fig. 6.22 Market vignette template. *Photoshop* on *Microsoft Surface Pro* using *Microsoft Surface Pen*. Makayla Lewis, 2019

Activity 6.2: Sketchnotes (Individual)
Learning objective—Create a sketchnote from a recording
Time—20 minutes
Materials—Basic materials needed, and any extra items that may be required (e.g. paper, pen, and a selection of markers)
Procedure:

- Find an informative video on the TedTalk platform.
- Start by understanding the content you want to sketchnote: is it a talk, lecture, book, user interview, etc.? Do you know the topic enough to sketchnote without preparation? If not, we recommend doing background research beforehand and creating relevant icons; this will help you later.
- Decide if your sketchnote will be a horizontal or vertical layout.
- Select a sketchnote layout style using Fig. 6.13, and roughly map the flow using a pencil.
- Write down the title at the top of the page. Use bold, clear lettering for this; you want it to be noticeable. You can also use decorative lettering to make it visually appealing.

6.6 Hands-On Activities

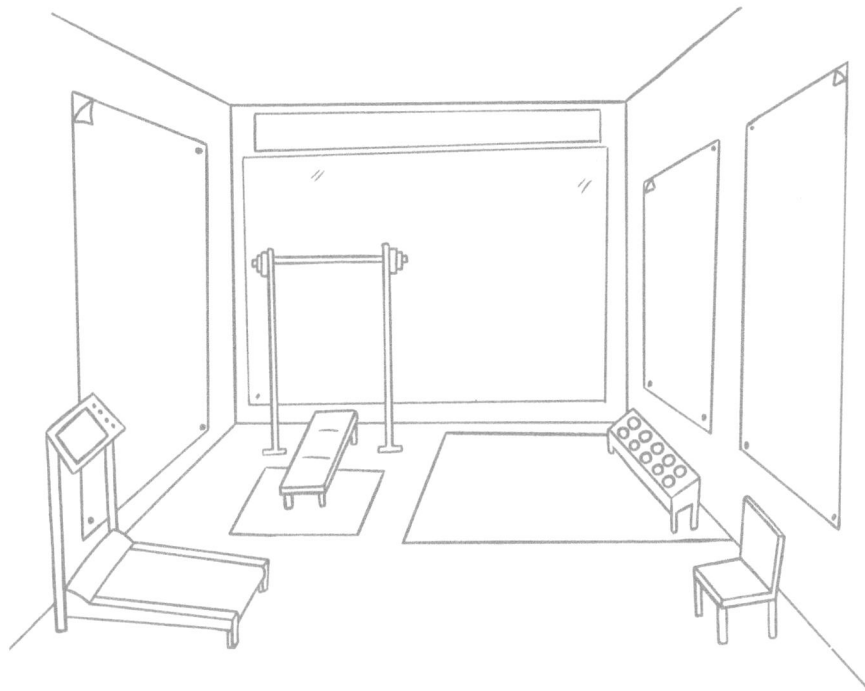

Fig. 6.23 Gym vignette template. *Photoshop* on *Microsoft Surface Pro* using *Microsoft Surface Pen*. Makayla Lewis, 2019

- Start the video:
  (a) Use icons and sketches to depict critical points or quotes.
  (b) Use connectors and separators to show relationships between different ideas. Remember, white space can provide separation and stop the sketchnote from being too busy.
  (c) Use different types of lettering to emphasise critical points.
  (d) Your sketchnote is a mix of sketches and notes; thus, you can write; we think new sketchnotes often need to remember this.
  (e) Use a marker (pastel shade) to highlight important text, concepts, or objectives.
- When the video ends, review and revise your sketchnote.
- Your first sketchnote will be challenging, find another video and try again. When you are confident, try at an event, e.g. an in-person lecture or an online meetup presentation, e.g. Figs. 6.24 and 6.25.

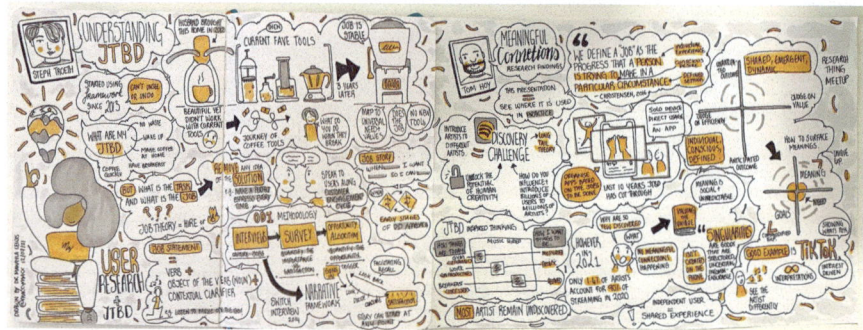

Fig. 6.24 Sketchnote from "User Research and Jobs to be Done" for *Research.Thing* meetup. Fineliner pen and marker on paper. Makayla Lewis, 2021

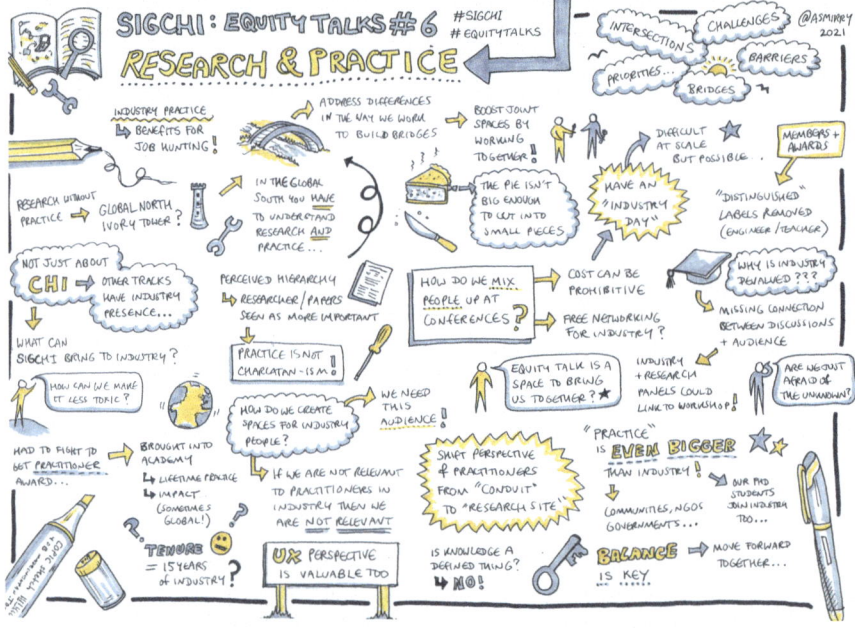

Fig. 6.25 Sketchnote from one of a series of "Equity Talks" for ACM SIGCHI—Research and Practice in HCI. Fineliner pen and marker on paper. Miriam Sturdee, 2021

Activity 6.3: Storyboards and Comics (Individual)
Learning objective—Quickly sketch storyboards, refine and reduce frame content, e.g. Figure 6.18
Time—15 minutes
Materials—Fineliner pen, A4 paper or sketchbook

6.6 Hands-On Activities

Procedure:

- Using the following scenario, sketch one frame for each bullet point:
    - Steve and Mary have taken their small dog Pip to the nearby park for a run. The park is quiet with a small forest area on the right, kids' playground on the left, and an open field in the centre.
    - Mary shouts, "Pip come, stay away from the kid's playground, get your stick, and come into the open field".
    - Fifteen minutes later, Steve turns to Mary—"I don't think this rain will let up; Pip soaked; maybe we should head home".
- You may wish to consider what the salient points are, the physical characteristics of the scene, and how to show the passage of time and change in weather.
- And just to keep you on your toes! Now reduce your narrative to ONLY ONE frame.

As humans we struggle to come up with ideas if we aren't given a prompt. If we frame a question as a "how might we" (Hanington & Martin, 2019), then it helps people frame a problem. This technique is often used in UX design. If you want to further this activity, or work as a group, ask yourself/the group: "how might we" support dog owners to better plan dog walks during the rainy season? Then brainstorm application or physical device features and sketch the interface (Figs. 6.26, 6.27 and 6.28).

Fig. 6.26 "Favourite pen" comic. Fineliner pen on paper. Makayla Lewis, 2020

Fig. 6.27 "Migraine" comic. Fineliner pen on paper. Makayla Lewis, 2020

Activity 6.4: Without Words (Group)
Learning objective—Sketch out understandable information about yourself without using text
Time—5–10 minutes sketching, 10 minutes annotating and discussing
Materials—Fineliners, A4 paper, Post-it notes, wall-safe removable adhesive
Procedure:

- As a group, choose either hobbies or research/project focus.
- Each person should sketch their hobby or focus (depending on the class choice) using imagery ONLY—no words.
- After everyone has finished their image, they should stick them to the wall or onto a whiteboard.

6.6 Hands-On Activities

Fig. 6.28 "Video games" comic. Fineliner pen on paper. Makayla Lewis, 2020

- Each member of the class should then take a block of Post-it notes, and for each image, write down what they THINK the person's hobby or focus is.
- When everyone has completed their Post-it, collectively they should stick their notes on or around the corresponding image.
- Each person then describes what was in their image in turn and who was correct.
- Discuss as a group how successful the sketches were and how they could be refined.

If using an online whiteboard, you can use this activity with remote classes (see the Without Words, Digital Edition in Chap. 10) or even as a fun warm-up for workshops! Examples by us both from an online course can be seen in Fig. 6.29. You can also switch what the "without words" topic is for variety, e.g. where you last went on holiday, favourite interactive device or technology, and so forth.

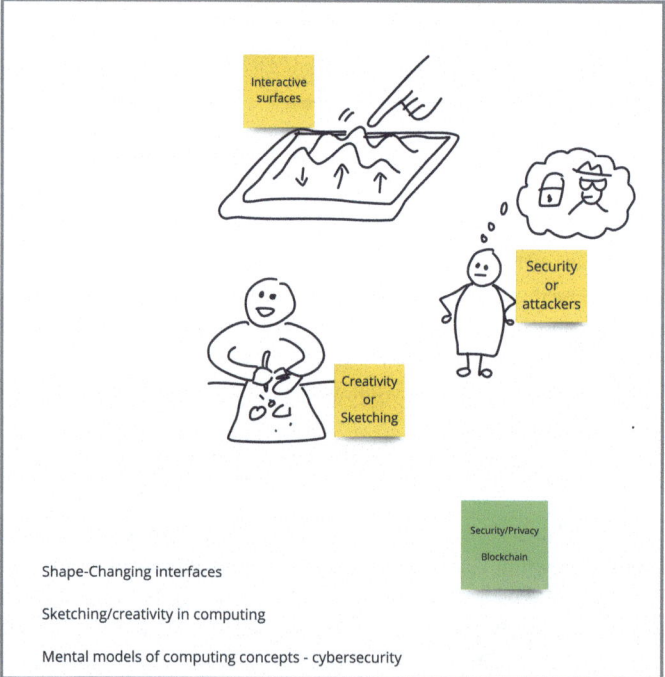

Fig. 6.29 Makayla and Miriam's "Without Words" responses for ACM CHI 2021 sketching in HCI course. *Miro* Online Whiteboard on *Apple iPad Pro* using *Apple Pencil*. Makayla Lewis and Miriam Sturdee, 2021

References

Books, Papers, and Articles

Disney Editions. 2008. Walt Disney animation studios – The archive series: Story. Disney Editions

Disney Editions. 2011a. Walt Disney animation studios – The archive series: Design. Disney Editions

Disney Editions. 2011b. Walt Disney animation studios – The archive series: Layout and background. Disney Editions

Eisner, W. (2008). *Comics and sequential art: Principles and practices from the legendary cartoonist*. WW Norton & Company.

Gibbons, D., & Pilcher, T. (2017). *How comics work*. Wellfleet Press.

Greenberg, S., Carpendale, S., Marquardt, N., & Buxton, B. (2012). *Sketching user experiences: The workbook*. Elsevier.

Hanington, B., & Martin, B. (2019). *Universal methods of design expanded and revised: 125 Ways to research complex problems, develop innovative ideas, and design effective solutions*. Rockport publishers.

Lengyel, D., Kaipainen, K., Sturdee, M., Heron, M., Lewis, M. and Liddle, J., 2023. Hands-on workshop on tabletop role-playing for inclusive design: Imagining sustainable futures for 'older adults'.

Lewis, M., Sturdee, M., Miers, J., Davis, J. U., & Hoang, T. (2022, April). Exploring AltNarrative in HCI imagery and comics. In *CHI conference on human factors in computing systems extended abstracts* (pp. 1–13).

McCloud, S., & Martin, M. (1993). *Understanding comics: The invisible art* (Vol. 106). Kitchen Sink Press.

Sturdee, M. (2023). Rage against the AI machine: A sketch in time. *Interactions, 30*(5), 8–9.

Sturdee, M., Coulton, P., Lindley, J. G., Stead, M., Ali, H., & Hudson-Smith, A. (2016, May). Design fiction: How to build a Voight-Kampff machine. In *Proceedings of the 2016 CHI conference extended abstracts on human factors in computing systems* (pp. 375–386).

Williams, R. (2012). *The animator's survival kit: A manual of methods, principles and formulas for classical, computer, games, stop motion and internet animators*. Macmillan.

Websites

W1 Oxford Online English Dictionary entry for 'vignette' – www.oed.com/search/dictionary/?scope=Entries&q=vignette

W2 Good example of a sketchnote journal – www.leuchtturm1917.co.uk/sketchnote-journal-english.html

W3 A collection of sketchnotes by Makayla Lewis – www.flickr.com/photos/makaylalewis/sets/72157633090981769

W4 A collection of sketchnotes from Mauro Toselli – www.maurotoselli.com/sketchnote-visuals

W5 A collection of sketchnotes from Mike Rohde – www.rohdesign.com/sketchnotes

W6 A collection of sketchnotes from Michael Clayton – https://www.profclayton.com/

Further Reading

Barry, L. (2019). Making comics. Drawn and Quarterly Illustrated Edition.

Dimeo, R. (2016). Sketchnoting: An analog skill in the digital age. *ACM Sigcas Computers and Society, 46*(3), 9–16.

Fox, T. (2022). *Fundamentals of character design: How to create engaging characters for illustration, animation & visual development.* 3dtotal Publishing.

Groth, G. (2020). Health care, disability, illness & comics. 2020. *The Comics Journal.* #305. Fantagraphics Books.

Heller, S. (2012). *Comic sketchbooks.* Thames and Hudson Ltd.

Lindberg, O.. Makayla Lewis on the power of sketchnoting in UX design. Adobe Xd. https://xd.adobe.com/ideas/perspectives/interviews/makayla-lewis-power-sketchnoting-ux-design/

Rohde, M. (2013). *The sketchnote handbook: The illustrated guide to visual notetaking.* Peachpit Press.

Rohde, M. (2014). *The sketchnote workbook: Advanced techniques for taking visual notes you can use anywhere.* Pearson Education.

Sturdee, M., Lewis, M., & Marquardt, N. (2018). SketchBlog# 1: The rise and rise of the sketchnote. *Interactions, 25*(6), 6–8.

Toselli, M. (2019). *The xLontras theory of sketchnote.* Independently Published.

Chapter 7
Design Fiction and Speculative Sketching

7.1 Sketching the Future!

By now you will be familiar with the ease with which we can create, ideate, iterate, and explore any number of topics and thoughts using the power of the pen and pencil. You've got a good grasp of lines, objects, people, and storytelling with visual narrative. So naturally now is the time to start *drawing things that do not exist*…

Don't panic! Yes, we ask you to observe the world around you and turn it into lines which communicate its properties directly, but you're now the master of real things; it is time to move on to future things. This might seem like a tall order—but actually it is even easier than drawing things that already exist, because YOU dictate what they look like and how they might behave in a world of your own creation.

We're not expecting you to sketch fantasy realms or space aliens (although both are lots of fun to imagine); instead, we ask you to turn your hand to sketching technology that could possibly or plausibly exist at some time in the future. This future could be in five, or a hundred, years, but the important thing is to sketch something that helps other people to imagine what it might be like to interact with that technology, and how it might work.

When we apply sketching to HCI in terms of prototype and application design, then this becomes part of a research process with a definite end goal or product, but if we spread our imagination beyond our immediate environment, timeline, and possibilities, then it might become something called design fiction.

By the end of this chapter, you should be able to:

1. Quickly brainstorm and sketch ideas for new technologies, interactions, and narratives.
2. Build up and sketch stories that show how these new ideas might be used.
3. Use your sketches to build a visual world around your novel ideas.

7.2 What Is Design Fiction and Speculative Design?

Design fiction has become popular in HCI research and is linked to our ability to imagine a future beyond our current technological means (Sterling, 2009; Lindley & Coulton, 2015). It enables us to work through problems and innovations from the near future and plausible to the far future and possible (Dunne & Raby, 2013). Within design fiction you will find methods that resonate with user experience design, such as physical and video prototyping, scenario generation, and storytelling—all available for a potential "user" to react to or be inspired by. In this context, it is also possible to consider the future of even mundane technology, to avoid potential pitfalls or react and produce adaptations quickly. So rarely in HCI do we consider the consequences of a developing and building single technology or digital innovation—and perhaps in many cases, it seems impossible to do so, but a healthy imagination and the ability to communicate ideas and concepts with sketching are great ways to help explore the future.

A widely accepted definition of design fiction is taken from Bruce Sterling, in his talk that is thought to be amongst the first mentions of the technique "Design Fiction is the deliberate use of diegetic prototypes to suspend disbelief about change" (Sterling, 2013). Diegetic loosely means that the "narrator"—or storyteller—presents the thoughts and feelings of the character (or world) to the intended audience. Design fiction has also developed since this original definition in that it has now become about "telling worlds, not just stories" (Coulton et al., 2017)—and it is these worlds that sketching is so perfect for, in ideating, elaborating, and presenting to others.

There is also overlap with other new disciplines such as speculative design (Dunne & Raby, 2013; see W2) and critical design, but the idea of forward facing, creative innovation remains at the core of futuring in all fields. As a point of comparison, speculative design artefacts are often provocative and focused on developing debate rather than "suspending disbelief" as we might find in design fiction. Regardless, artefact generation is key to helping others understand the world which we aim toward, and sketching can help us achieve this.

7.3 Speculative Sketching for Ideation and Exploration

When we sketch through ideas, we are usually always speculating. We think, we sketch, we adapt and iterate—aiming for something that makes sense. We usually do this for projects with immediate application; however, we already have a goal and an idea of the design we are going for. With speculative sketching, we are future proofing our work. If we can discover the flaws in our project plan before we embark on what could be a time-intensive and costly project, then we can pivot to new ideas. In this way sketching can be a valuable aid to the early design process—even for physical prototypes! In this context, we see speculative sketching as part of the

7.3 Speculative Sketching for Ideation and Exploration

futuring, design fiction, or speculative design process, but it can also be used for projects where the outputs are immediately usable or plausible to build.

Figure 7.1 shows worked up participant ideas from a research study with members of the public on actuated shape-changing interfaces to deduce its efficacy and plausibility as a research direction (Sturdee et al. 2015). Shape-changing interfaces are a novel area of research in HCI that look at actuated (moving and movable) 3D physical interfaces, covering everything from foldable devices right through to nano-bot and swarm interfaces. Many prototypes in this area are lo-fi or extremely large and complex to build and maintain, making it important to only BUILD what you need—and therefore, we can SKETCH what we need to envisage the possibilities of these interfaces. This technique will, of course, work for other types of novel technology and interaction.

The original prompts for the shape-changing key idea from participants (all individually ideated, then written or sketched) can be seen in Fig. 7.2. The elaborative sketch looks into different technologies for both the key and lock mechanism, based on current, existing technology, and imagines the inner workings of both. After the initial sketching, colour was added to better highlight the elements for presentation. Annotations show the thought process involved alongside each component sketch.

The complexity of this key actually is its undoing—the amount of moving parts involved and the potential weight of components make it a nonstarter—this is a case of a solution looking for a problem! With the widespread use of digitally encoded keys, this kind of device becomes even less useful, as it needs to use both physical and digital technology to actualise it and therefore is more effort to build and deploy.

A second example (Fig. 7.3) shows another popular idea from the same workshop—using shape-changing interfaces for gaming. Again, the participant ideas are

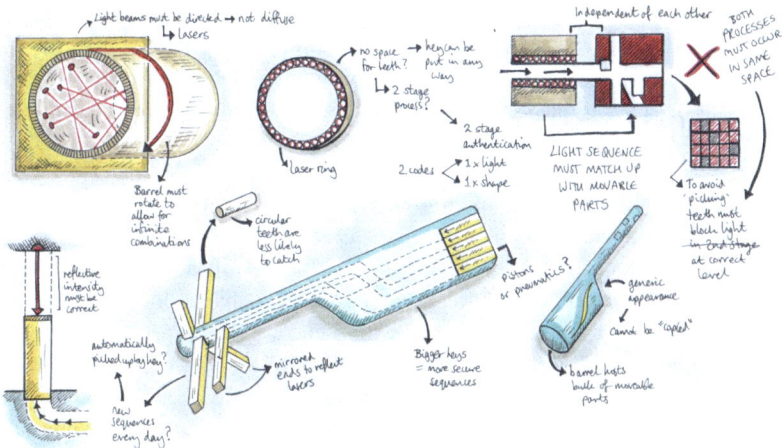

Fig. 7.1 Speculative sketching around the idea of shape-changing keys—plausible (we could imagine using this in day-to-day life), and near future (i.e. we could build it either now or in the next few years based on current technology). Fineliner pen and marker on paper. Miriam Sturdee, 2015

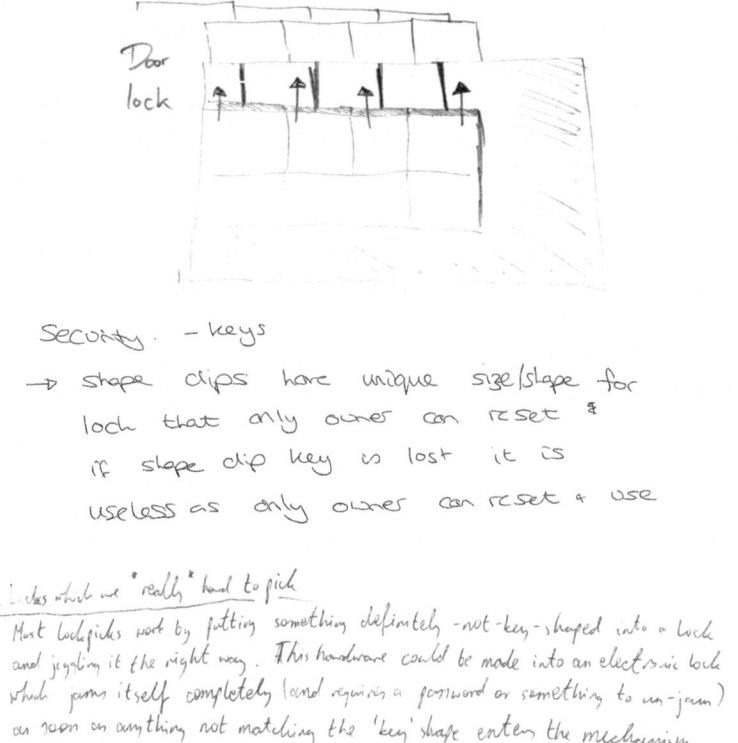

Fig. 7.2 Original participant ideas that prompted the sketching exploration of shape-changing keys. Pencil on paper. Anonymous participant, 2015 (from Sturdee, 2015)

shown beneath (Fig. 7.4). The problem with gaming using this novel technology is primarily that it is a very physical technology, and to create a fully immersive experience requires a lot of mass and energy. The use of actuators in this case means that their length requires vast storage, and their physicality could cause actual bodily harm. If we remove the need for physical interaction, then the 3D effect becomes passive and then requires sensors to register the various interactions, meaning the user must wear a gaming suit to register each game event that causes contact. Further, if this level of haptic detachment is necessary, it is no different from building wearables to help engage with VR. Sorry actuated shape-changing interfaces—we need nano-bots! But they also come with their own issues of mass and potential harm and are a lot further off becoming commercialised—making gaming with shape-changing interfaces a sketch further into the future…

Finally, speculative sketching can also make use of found or everyday objects. Futuring and sketching just need a little inspiration. Figure 7.5 shows Makayla's participation in an online workshop on the future of surveillance, supported by objects in the home, combined with sketching. The starting-off point for sketching

7.3 Speculative Sketching for Ideation and Exploration

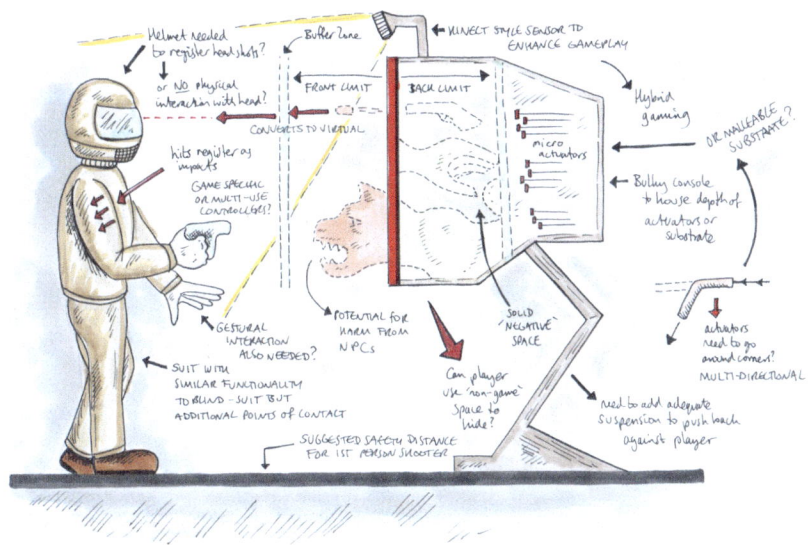

Fig. 7.3 An actuated shape-changing game sketch, examining how we could build the hardware and how we might interact—possible? Further future. Fineliner pen and marker on paper. Miriam Sturdee, 2015

the future can be anything, as long as it makes you think carefully through the problem (or future problem) at hand. Examples can be very specific (as in the case studies on shape-changing interfaces) or broad (wider societal and technological issues). Sketching can support all levels of futuring.

Practical Application Tips

- Keep your style loose and don't fixate on a particular scene or layout at first—this will emerge organically.
- Don't be afraid to cross things out or scribble over mistakes; this is visual thinking, not presentation—if you want to make a version for presentation later, rework your sketches and notes.
- Annotation is your friend; you are dealing with things that do not exist, and they may be hard to make sense of!
- If you find yourself going down a "rabbit hole", try to follow it to its conclusion; use another piece of paper if you need to.
- If the idea doesn't work in your sketch, then it probably won't work in real life. Know when to change your approach…
- If sketching broader concepts, think big; and identify the audience first.
- Icons can be used to depict future concepts, much as they can for current ones; building a future icon library can support innovation.

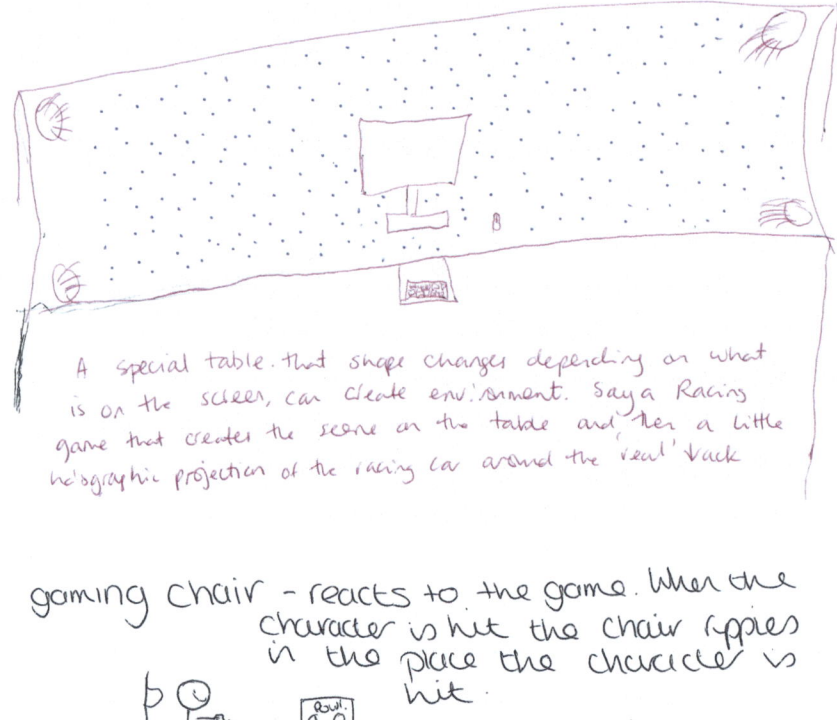

A special table that shape changes depending on what is on the screen, can create environment. Say a Racing game that creates the scene on the table and then a little holographic projection of the racing car around the 'real' track

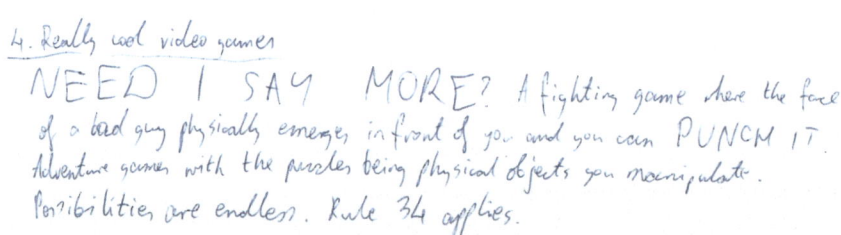

gaming chair - reacts to the game. When the character is hit the chair ripples in the place the character is hit.
- Character hit in leg ⇒ player feels a bump in chair by their leg.

4. Really cool video games
NEED I SAY MORE? A fighting game where the face of a bad guy physically emerges in front of you and you can PUNCH IT. Adventure games with the puzzles being physical objects you manipulate. Possibilities are endless. Rule 34 applies.

Fig. 7.4 Participant ideation and sketching around the premise of shape-changing interfaces being used for games. Pen on paper. Anonymous participant, 2015 (from Sturdee, 2015)

7.4 Speculative Sketching for Visual Narratives

Speculative sketching can also cover visual narratives and scenarios based on futuristic ideas. These differ from the exploratory sketches in Figs. 7.1 and 7.2, as they are not figuring out the plausibility of the build process, but telling stories about how

7.4 Speculative Sketching for Visual Narratives

Fig. 7.5 What do we think the future of surveillance will look like? Hybrid analogue/digital online activity. Fineliner pen on Post-it notes and desk objects. Makayla Lewis, 2020

a novel technology might be used. In these cases, the focus is on the human interaction with the device and encourages the viewer to think about the potential adoption of certain types of technology and raises issues that feed back into the current zeitgeist as well as future possibilities.

Figure 7.6 is again building on the ideas of others but in this case examines the claims made in HCI research papers about "future work" (Sturdee & Lindley, 2019). This particular paper imagines that actuated "hair" or fur could be used when designing robotic pets. This fur would have the potential to both be a display and also be soft and comforting to touch. The scenario shown imagines a robotic "helper" for an elderly person with limited mobility who lives by themself. It acts as a virtual doorbell camera, carries objects, and is Wi-Fi connected to be used as a

Fig. 7.6 Finishing the story, a future scenario featuring a robotic pet based on a "future work" section of a research paper (in Sturdee & Lindley, 2019). Fineliner pen and marker on paper. Miriam Sturdee, 2017

tablet device when necessary. If that is not enough, it can also become a cupholder side table. Far-fetched? Useful? Dangerous?

Again, even though we do not think explicitly about the hardware, the *context* of use here allows us to interrogate the idea. The robotic pet has agency; it can react in humanlike expected ways to be a useful assistant. It is a screen, a conversational user interface (CUI), a robot, and a pet. Although the technology we would need to actualise this device is not yet fully available, the story here resonates with current issues of privacy and security around technology (such as the children's doll which was found to be spying on people) (see W1). What if the pet was hacked, what damage could it do?

Sketching in this context serves as both a repository of possible future technology and as a talking point, a boundary object to connect issues with technology to easy-to-understand content that could be used to communicate not only with experts in HCI but members of the public. Sketching makes it easier to break down barriers between people and allows the researcher to better connect with others—far removed from the feeling of being filmed or recorded.

Practical Application Tips

- Use the tips and activities from Chap. 6 to help you design your future visual narratives; it's the same thing, just further into the future!
- Incidental details in the narrative help situate the technology in everyday life, for example, small background objects, or door signage.
- Use highlighting to centre the future technology depicted in the story.
- Try and explore a range of possible interactions rather than focusing on one particular feature of the technology.

7.5 Speculative Sketching for Artefacts and World Building

Many projects in design fiction make or build artefacts, everything from digital WOZ (Wizard of Oz) forks and canes (see the film *Uninvited Guests* by SuperFlux Lab in the website list! W3) to Ikea catalogues of the future (from Near Future Laboratory, see W4). Artefacts can provide a central focus around which you can then build other aspects of the "world" it lives in, and they don't have to be physical artefacts, or even fully digital. Of course sketching has a part to play too.

Figure 7.7 shows the front cover and two internal pages of a computer game manual for a shape-changing interface game called *First Hand* (Sturdee et al., 2017). This digital (but printable) artefact mimics the style of Super Nintendo game guides from the 1990s and is a full game manual of 38 pages, including contents and front and back covers. Whilst the cover image is certainly guilty of being "art not ideas" (and it started with a sketch!), the images inside the booklet are almost entirely quick, rough sketches which have then been digitally coloured. The surrounding context means that their "sketchiness" is not detracting from the message and helps the reader imagine the gameplay of *First Hand*. The full game manual is

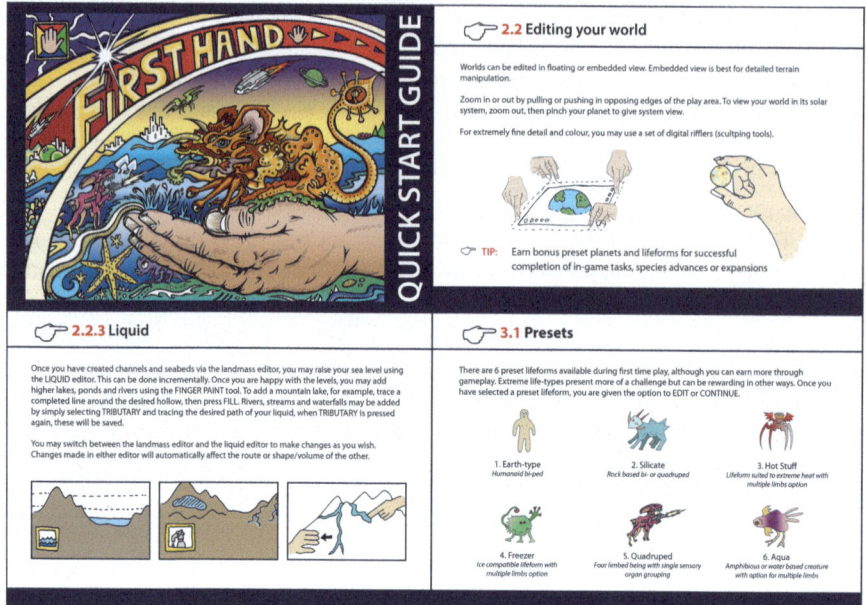

Fig. 7.7 Game manual cover and two example pages for the shape-changing interface game, *First Hand*. Fineliner sketch and digital colour in *Photoshop* (in Sturdee et al., 2017). Miriam Sturdee, 2017

similar (in part) to the speculative sketches imagining shape-changing keys and games, except that the aspect of "plausibility" has been abandoned in favour of setting the scene to show that the technology and hardware issues have been solved. "Working through the idea" has been expanded to encompass not only how to interact with the hardware itself but also how the entire game mechanics are realised, from single to multiplayer online mode.

The world in which the artefact game manual exists then is furthered with the development of an accompanying magazine article (Fig. 7.8). The article itself exists in a fictional magazine, in a fictional magazine section, and contains mocked up game interactions (a mix of photographs and sketches) and even a small section of weekly cartoons by a fictional artist. The article is also fully readable and written in the style of a general interest weekly magazine that could plausibly be part of a "weekend newspaper".

Including topical references (such as *Pokemon Go* in this case), even though the artefact is future facing, can help the viewer to "suspend disbelief" (see Fig. 7.9). Every mundane feature can contribute to the world-building effect! These small sketches add that extra layer to the world in which *First Hand* exists.

A sketch may not be shiny and fancy, but it has the ability to transport us into future worlds and think beyond the technologies we already interact with. It can also be used in a hybrid manner to enhance digital materials as place information into high-fidelity imagery (see Fig. 7.10). Try speculative sketching for yourself and build some new worlds by engaging with the activities in the next section.

7.5 Speculative Sketching for Artefacts and World Building 135

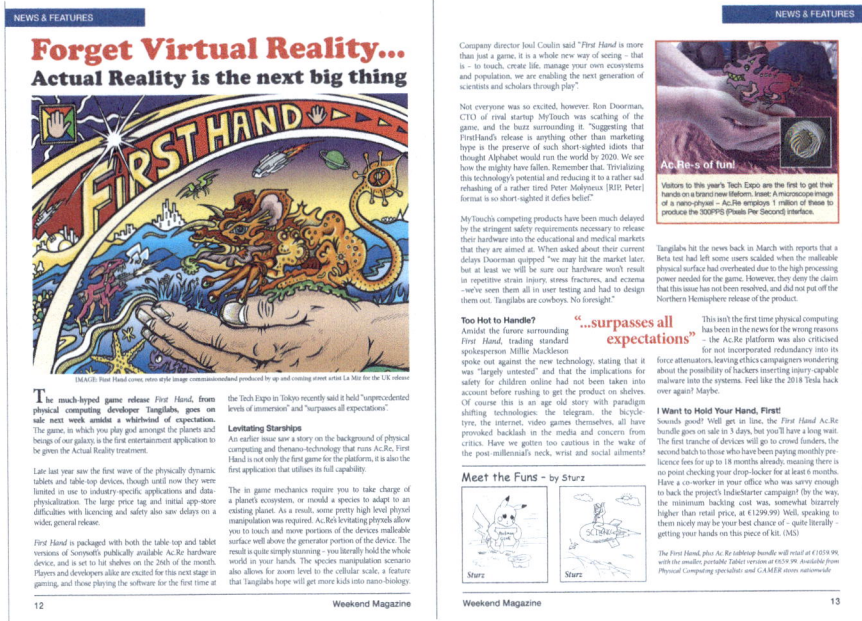

Fig. 7.8 Fictional magazine article and layout describing the launch event of the game, *First Hand*. Digital construct. Miriam Sturdee, 2017

Fig. 7.9 Detail from the article on shape-changing games, Pikachu cartoon by fictional comic artist Sturz. Fineliner pen. Miriam Sturdee 2017

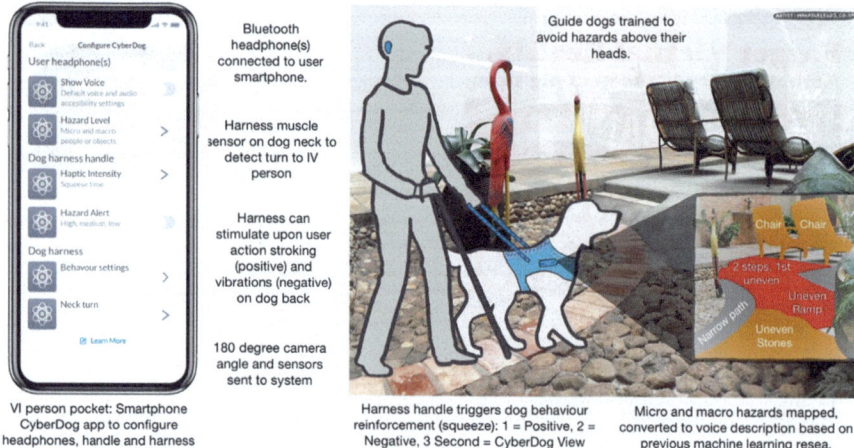

Fig. 7.10 Sketching and forward thinking for other people, future engagement for blind people. *Photoshop* on *Microsoft Surface Pro* using *Microsoft Surface Pen*. Makayla Lewis, 2018

Practical Application Tips

- Try to develop a range of artefacts and stories to add depth to your fictional world.
- Mundane details can really help a world "exist"—after all mundanity is reality.
- Mixed media is your friend with world building, try combining sketches with other approaches.
- Fine details help pull the viewer deeper into the world and can encourage engagement with sceptical audiences.
- Adding elements of humour has the same effect as mundanity—after all, these are some of the things that make us Human.

7.6 Hands-On Activities

Are you ready to sketch the future? Or even a galaxy far, far away?

Activity 7.1: Sketching Fictions (Individual Activity)
Learning objective—Analysis and sketching of future objects to assess plausibility and technology needed
Time—20 minutes
Materials—Fineliner, paper or sketchbook, coloured pens or pencils as needed
Procedure:

- Think of a futuristic book you have read or film/tv series you have watched recently. If you have not engaged with science fiction, then go online and search for short science fiction stories for ideas.
- Pick one future technology that the story uses.

7.6 Hands-On Activities

- Sketch what you think that object looks like from its description (if from a book) or sketch the visual depiction of the item from a film from memory or from Internet resources.
- Consider what material it is made from.
- Think about whether it has any moving parts.
- How much might it weigh, or is it weightless?
- What existing technology does it use, what future or unknown technology does it employ?
- What would we need to achieve to build it?
- Sketch and annotate your answers to these questions and other questions that you might ask yourself during the process.

For this activity, the idea is to dissect a future fictional idea and realise how plausible it might be and how far in the future it might exist. Could we build it in 50 years? 500 years? Is it simply "magical" in its construction? Did you have to sketch abstract concepts and how did you approach this? This activity employs a form of "reverse world building" for design fiction—the world in which the technology and its stories already exist, you are adding detail to that world.

Activity 7.2: Objects of the Future (Pair Activity)
Learning objective—Ideate around product development and future use cases using sketching
Time—30 minutes (3 rounds of 6–8 minutes, prep and discussion)
Materials—Fineliner, paper or sketchbook, coloured pens or pencils
Procedure:

- Each member of the pair should think of an everyday, mundane technology and write it down at the top of an A4 sheet of paper.
- Swap with the other person.
- For the technology you have been given, you are going to be redesigning it for 50 years in the future.
- Sketch out the technology as it is now, and then, using a different colour, start adapting it and annotating it. Add a brief context and scenario of use. Consider: What has changed in 50 years? Do we have different requirements? Is this technology now obsolete? If so, what replaced it?
- After 6–8 minutes, swap your sketch with the person who gave you the prompt.
- You are now thinking 150 years in the future. Using a different colour again, the sketcher should build on the ideas of their partner, iterate, and change again. Consider: Is the technology still necessary? Has it changed size or type of technology? Are the users the same people?
- After 6–8 minutes, swap your sketch with the person who gave you the prompt.
- Now you are 500 years in the future. Using a different colour again, the sketcher should build on the ideas of their partner, iterate, and change again. It is ok at this point if the ideas become more far-fetched, as we are dealing with ideas that could be thought of as impossible!

- After this final round, discuss your sketches with your partner, and consider the trajectory of the technology. If the technology employs novel ideas and hardware or software, relate this back to current science. At what point did the thinking become "magical"? At what point might this technology become obsolete?

By futuring existing technology, we may spring upon useful ideas and contexts of use which we can actually employ or develop in a more immediate timeline. If the ideas seem far-fetched, then they could be aspirational! Or you could represent them as abstract concepts (Figs. 7.9 and 7.11). Only a few hundred years ago, the motor car would have been a huge leap of faith—like the famous (yet allegedly untrue—Vlaskovits, 2011) quote by Thomas Ford, "If I had asked people what they wanted, they would have said faster horses". The factuality of the quote here is unimportant; however, what IS important is the ability of humans to think ahead and innovate, and by sketching the future, you are doing just that. Now get out there and design those space horses.

Activity 7.3: HCI Improv (Group Activity)
Learning objective—Work in a team using sketching to co-create, diagram, and design new technologies and visual narratives
Time—30 minutes
Materials—Post-it notes, A4 paper, A3 paper or flipchart, fineliner pens, thicker marker pens, coloured markers as needed

Fig. 7.11 "Sketch your post-COVID future". *Photoshop* on *Microsoft Surface Pro* using *Microsoft Surface Pen*. Makayla Lewis, 2020

7.6 Hands-On Activities

Procedure:

- One person should facilitate the group (either a lecturer, teaching assistant, or volunteer student).
- The facilitator should ask the group to shout out spontaneous ideas for a type of existing (or at least plausible) general type of technology, e.g. chest mounted wearables, virtual reality, shape-changing interfaces. These ideas should not be contextualised. Each idea should be written down as they are shouted out (or loudly spoken!).
- Next, the facilitator should ask for user group suggestions, for example, the elderly, dogs, the police, chefs, and so forth, and write these down too. The group should not over think this, or try to immediately link it back to the first ideas.
- Finally, the facilitator should ask for random contexts or scenarios of use, e.g. in the mountains, whilst travelling by air, at high schools. The stranger the better!
- Once there are several ideas for each part, either the group can vote, or the chosen facilitator can decide at random. This actually works better if each idea does not obviously link up with the others.
- Now it is time for HCI Improv! Students should form groups of four or five people.
- First, the group should spend 5 minutes brainstorming ideas for specific iterations of the chosen technology that might fit the user and use case. This should be quick fire, and group members should write these down on Post-it notes and stick these in a group onto a table, wall, or large piece of paper. After 5 minutes, the group should vote on one ideas to take forward to the next step.
- The chosen idea should now be sketched speculatively, diagrammed, and annotated as a group. You can use Post-it notes to facilitate building up the idea or nominate a couple of group members to lead on sketching this part. Spend up to 10 minutes on this step.
- The final stage is to develop a story about your technology and its user! Using the techniques you developed in Chap. 6, develop a visual narrative that shows the world this technology exists in and how the user might utilise it. For this part, Post-it notes allow each group member to create one or two frames of the visual narrative, using the diagram as a common resource, and discuss the overarching story before committing frames. However, the Post-it storyboard frames can be easily replaced or swapped with other team members to elaborate upon.
- All groups should then present and "pitch" their idea to the room, and the facilitator can choose a "winner", or the group can vote on their favourite.

We have been running HCI Improv since 2016 and have touched upon topics as far reaching as chest-mounted wearables for children hiking in the mountains, digital cat litter trays for air travel, and conversational user interfaces for pirates at a conference (Fig. 7.12). Sometimes the sillier the context and use case, the more inventive and fun the activity can be—but equally you could pre-load the technology, use case, and user, as a more structured class task, which we sometimes do when time is short or the audience has specific knowledge, or the potential results are applicable to their field of study.

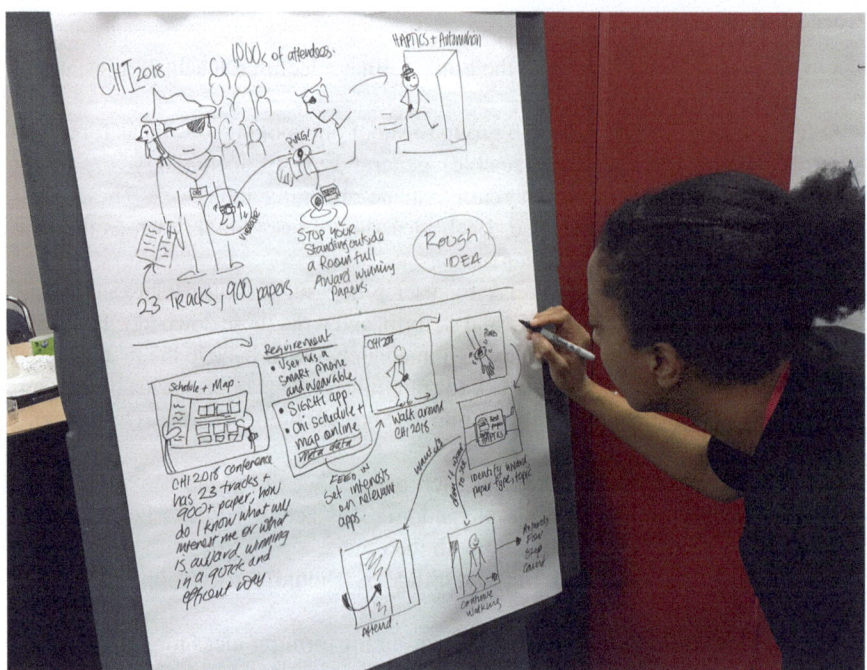

Fig. 7.12 Makayla sketches HCI Improv ideas (pirate wearables). Sketching in HCI Course (Lewis et al., 2018). Marker on flipchart.

References

Books and Papers

Coulton, P., Lindley, J.G., Sturdee, M., & Stead, M. (2017). *Design fiction as world building.* In Proceedings of the 3nd Biennial Research Through Design Conference (pp. 1–16).

Dunne, A., & Raby, F. (2013). *Speculative everything: Design, fiction, and social dreaming.* MIT Press.

Lewis, M., Sturdee, M., & Marquardt, N. (2018, April). Applied sketching in HCI: Hands-on course of sketching techniques. In *Extended abstracts of the 2018 CHI conference on human factors in computing systems* (pp. 1–4).

Lindley, J., & Coulton, P. (2015, July). Back to the future: 10 years of design fiction. In *Proceedings of the 2015 British HCI conference* (pp. 210–211).

Sterling, B. (2009). *Cover story design fiction. interactions, 16*(3), 20–24.

Sterling, B. (2013). Fantasy prototypes and real disruption. *Keynote NEXT Berlin, 2013.* http://www.youtube.com/watch?v=2VIoRYPZk68

Sturdee, M., & Lindley, J. (2019, November). Sketching & drawing as future inquiry in HCI. In *Proceedings of the halfway to the future symposium 2019* (pp. 1–10).

Sturdee, M., Hardy, J., Dunn, N., & Alexander, J. (2015, November). A public ideation of shape-changing applications. In *Proceedings of the 2015 international conference on interactive tabletops & surfaces* (pp. 219–228).

Sturdee, M., Coulton, P., & Alexander, J. (2017). Using design fiction to inform shape-changing interface design and use. *The Design Journal, 20*(sup1), S4146–S4157.

Vlaskovits, P. (2011). Henry Ford, innovation, and that "faster horse" quote. *Harvard Business Review, 29*(08), 2011.

Websites

W1 Article about the possible spying doll – www.nytimes.com/2017/02/17/technology/cayla-talking-doll-hackers.html
W2 Homepage of the speculative design studio of Dunne & Raby www.dunneandraby.co.uk
W3 Uninvited Guests by Superflux Lab www.superflux.in/index.php/work/uninvited-guests/#
W4 Near Future Laboratory Ikea Catalogue www.ikea.nearfuturelaboratory.com/

Further Reading

Dunne, A., & Raby, F. (2013). *Speculative everything: Design, fiction, and social dreaming.* MIT Press.

Lindley, J., & Coulton, P. (2016, May). Pushing the limits of design fiction: The case for fictional research papers. In *Proceedings of the 2016 CHI conference on human factors in computing systems* (pp. 4032–4043).

Chapter 8
Accessibility of Sketches

8.1 Introduction

"The power of the web is in its universality. Access by everyone regardless of disability is essential." (Tim Berners-Lee, W1)

Remember that disabilities, impairments, and conditions can significantly impact how people interact with the world. In 1985, Ronald L. Mace proposed the idea of universal design:

"The design of products and environments to be usable by all people, to the greatest extent possible, without the need for adaptation or specialised design." (Ronald L. Mace, W2)

It is widely accepted in HCI that usability lies in our interaction with a application, product, service, and environment, which is measured by observing performance, satisfaction, and acceptability. To this end, there is an overlap between usability, user experience, and accessibility, the latter defined as "people can do what they need to do in a similar amount of time and effort as someone that does not have a disability" (W3). Accessibility designates that digital products, services, and environments we create are perceivable, understandable, operable, robust, and encompass the range of human diversity, i.e. usable experience for all. This chapter aims to provide insight and concrete techniques to improve the accessibility of your sketches.

By the end of this chapter, you should be able to:

1. Understand that accessible sketches are essential.
2. Identify different types of disabilities and the actions you can take to make your sketches accessible to people with specific disabilities.
3. Practise making your sketches accessible using sketches you have created previously.

8.2 Rationalisation

In recent years, the requirement or encouragement for social media users to add alternative texts to imagery (e.g. *X* (W4) and *Instagram* (W5)) and captions for short (e.g. *TikTok* (W6)) and long (e.g. *YouTube* (W7)) video content with audio has appeared across social media platforms. Thus, the importance of accessible digital content is being seen by researchers, designers, and developers, who are, in turn, distilling this requirement to their users. It can be argued that this action is linked to accessibility laws, regulations, guidelines, and standards in many countries that require digital content, including digital imagery (which you are creating), to be accessible for all.

Accessibility of digital content is not only a legal (in many countries) which does not mean you should not ensure your sketches are not accessible. It can allow digital content creators, in this instance, those who create imagery (sketchers such as yourselves), to reach a broader audience and demonstrate ethical considerations. The wider audience includes people with disabilities, older people with age-related disabilities, conditions, and impairments, and those using assistive technologies (ATs). Ensuring your sketches are well-structured, and appropriately described makes digital sketches (scanned/photographed or digitally drawn) more understandable and engaging for all, not just people with disabilities, a fundamental aspect of digital visual content creation.

From a research perspective, accessible sketches of your work shared on websites, social media platforms, and other digital communication tools / apps can improve search engine optimisation (SEO). This is due to search engines like *Google* and *Edge* using text descriptions to understand digital image content. This can lead to a greater reach (audience) e.g. when included on a university research webpage, open-access paper, or students' teaching and learning platform. Finally, we believe, as technology evolves and new technologies emerge, accessible content will become more critical, thus ensuring our sketches are accessible at the time of creation / sharing will act as "future proofing".

8.2.1 Sketching in HCI Accessibility

In HCI, the pursuit of accessibility is of increasing importance; our research, teaching, and outputs should be accessible to all. Sketching in HCI is no different; as you now know it is a powerful tool to translate empathy, experience, and knowledge of diverse user needs, preferences, and experiences. For this to occur, viewers needs to be able to understand, reflect, and discuss your sketches throughout the design process. Often, accessibility of our sketches is restricted to including diversity in the process, output, and discussions—which is brilliant. However, what about the accessibility of the sketches themselves? As HCI sketchers, we often do not understand how to make our sketches accessible.

We hope this reasoning has convinced you to ensure your sketches, which will be shared with others, are accessible.

8.3 Sketch and Disability

"1 billion people with disabilities. Everyone should be able to access and enjoy the web."
GAAD (W8)

Disability is a broad term for physical, sensory, cognitive, and mental health conditions. The United Kingdom government defines disability:

"A person is disabled under the Equality Act 2010 if they have a physical or mental impairment that has a 'substantial' and 'long-term' negative effect on ability to do normal daily activities." (W9)

Disabilities, conditions, and impairments can be congruential or acquired, temporary or permanent, and can vary in severity. People with disabilities may require accommodations or support to interact with digital products, services, and environments as those without disabilities—this includes the sketches you create.

There are a variety of disabilities, conditions, and impairments to consider when you choose to share your sketches with others; these include (but are not limited to) the following:

- **Blind and low vision**—sketchers and viewers may have difficulty perceiving your sketches use of colour. They may use alternative methods, such as alternative text descriptions via screen readers. Therefore, including AltText (and, where necessary, AltNarrative, discussed in Chaps. 7 and 8), appropriate colour contrast, clear structure (ensuring read order is consistent and logical and minimises horizontal engagement), and not using colour to convey information or meaning could be beneficial.
- **Physical**—sketchers and viewers may have difficulty interacting with sketches due to no or reduced motor function, they may experience tiredness when navigating large amounts of visual content, and/or require assistive technologies. Thus, including AltText, clear and well-structured visuals, and consistent and logical content chunking can be beneficial.
- **Cognitive**—sketchers and viewers, such as people with dyslexia or cognitive processing disabilities, may have difficulty understanding complex and/or abstract sketches; thus, we recommended ensuring your sketches are simple and well-structured thus mak the content/message more accessible.
- **Neurological**—sketchers and viewers with neurological conditions, such as epilepsy, may be affected by specific visual patterns in your sketches. Thus, the inclusion of geometric patterns with high contrasts of light and dark, such as stripes or bars, is discouraged.
- **Sensory processing**—sketchers and viewers, such as people on the autism spectrum, may be affected by visual stimuli. Thus, sketches accompanied by AltText that have good colour contrast, include annotations, and are simple with minimal distractions are encouraged.

- **Language**—sketchers and viewers with limited proficiency in the language may need help understanding your sketches' content and/or structure. Thus, considering the sketchers and viewers, you may need to take a multilingual approach (this will be discussed in Chap. 9). It is encouraged that using clear and universally understandable icons in your sketches (see Chap. 3) can make your sketch elements more accessible, e.g. W10.
- **Age related**—as we age, sketchers and viewers may experience declining vision and cognitive abilities (description and recommendations discussed above).

We encourage you to test the sketches you intend to share with our intended audience (viewers) to ensure it is appropriate for their needs (discussed further in Chap. 12). To explore disabilities, impairments, and conditions, we recommend reading W11, W12, and W13.

8.4 Additional Techniques for Accessible Sketches

8.4.1 Alternative Text

Image text description, referred to as alternative text (AltText), is a descriptive text that provides an alternative description for an image for people who cannot perceive visual content (in this instance, your sketches). You must create AltText that is clear and supports your viewers to engage with your sketches accurately, the critical elements of AltText should include the following:

Goal The first step is to identify the purpose of AltText. To help you with this, we recommend answering the following questions:

- Who is the viewer of your sketch? (This will determine the language level of the AltText)
- Is the sketch a simple icon or a complex visual telling a story, e.g. comic or storyboard? (This will help you to decide how detailed your text description should be)
- What is the key plot point (focus) in the sketch that you would like to highlight? (Creating a bulleted list can be helpful)

Write The second step is creating the text description; we recommend creating it in a word process or so you can check spelling and grammar before embedding it in the image. As per guidelines (W14), ensure your AltText is:

- Accurate and equivalent.
- Succinct.
- Not redundant.

The purpose of AltText is to present the same content and function of your sketch in text form, thus ensuring your descriptive text is clear, concise, accurate, and

8.4 Additional Techniques for Accessible Sketches

consistent. Do not use jargon, informal language, or symbols, as the sketchers and viewers reading or listening to your AltText may need help understanding the story you are trying to convey. If your sketch is complex and contains dialogue or data, we recommend providing an overview followed by a more detailed description, e.g. for a vignette or storyboard, we recommend adding "direction" (flow) prompts to your AltText, e.g. "leads to", "returns to…", "adds to", etc. This will ensure the narrative order is not lost.

For example (see Fig. 8.1), the AltText could be "A line sketch of a large full strawberry jam jar".

For scenes, you will need to be more detailed; for example (see Fig. 8.2), the AltText could be "Two figures which look like wooden ergonomic models are engaged in a martial arts demonstration. The figure on the left is standing ready to fight, movement lines by the hands showing the tension and readiness of the person. The person on the right is already raising their leg to kick out, arms outstretched for balance. The background shows a domed floor with a contrasting marker-sketched background. Between the two figures is some text stating "live demo", which is underlined twice".

Embed In HTML, the image attribute added to the image (<img>) as an attribute (specifies an alternative area that contains text):

<img src="img_person.jpg" alt="tall person in a jacket" width="500" height="600">

Fig. 8.1 Sketch of a strawberry jam jar. *Procreate* App on *Apple iPad Pro* using *Apple Pencil*. Makayla Lewis

Fig. 8.2 Sketch of two figures in black fineliner and coloured marker performing a martial arts live demonstration. Fineliner marker on paper. Miriam Sturdee, 2016

However, AltText can also be added to a PDF (W14), *Microsoft Word* and *Microsoft PowerPoint* (W15), and social media platforms; see Figs. 8.3 and 8.4 for examples, and software / applications often provide AltText support pages with instructions. Some tools do offer an option to automatically generate AltText using artificial intelligence (*Springer Nature* did so for this book); if you choose to use this, we recommend comparing the output with your sketches and make adjustments (edits) will likely be required (note: we manually edited all artificial intelligence generated AltText for this book).

As per *W3C* (W16), if you intend to add AltText via HTML, you should describe the sketch if it contains complex information, explain where the link will go if the sketch is inside an <a> element, and use alt="" If the image is only for decoration. Please remember not to use "image of ..." or "graphic of ..." or "sketch of" at the beginning of your AltText, as this is redundant for viewers who use a screen reader. For example, for Fig. 8.1:

<img src="img_LargeJamJar.jpg" alt="A line sketch of a large strawberry jam jar" width="500" height="600">

8.4 Additional Techniques for Accessible Sketches

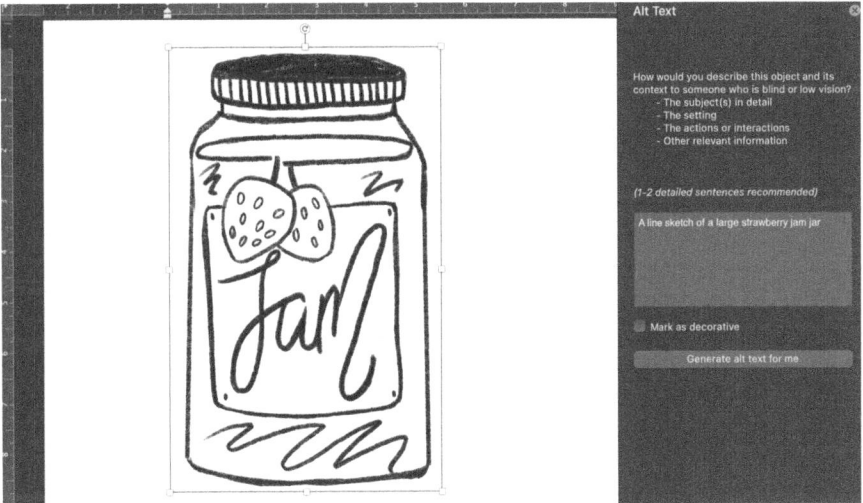

Fig. 8.3 *Microsoft Word* dialogue box to add AltText to an image in a research paper. Screenshot. 2023 (see W15)

Fig. 8.4 "*X*" form to add an image before posting. Screenshot. 2023 (see W4)

You probably noticed that all formatting present in AltText, e.g. paragraphs, bullet lists, etc., have been removed. AltText attributes only support plain text; thus, please remember this when constructing your text descriptions.

8.4.2 Alternative: AltNarrative

What do we mean by AltNarrative (Lewis et al., 2022)?

As visual researchers and practitioners with an interest in accessibility, Makayla, a screen reader user, shared with their collaborators, when we create beautiful, comic dialogues, the traditional view of AltText does not do the work justice nor convey accurately what the images depict. So we decided to create a publication that had no visible text, only visual narrative, with the story embedded entirely within AltText (Lewis et al., 2022). We wove together detailed image descriptions with flowing prose and created the idea of AltNarratives. AltNarrative tries to instil the same sense of creativity that we find in carefully constructed sketches, within audio text for screen readers, thus creating a engaging story even if the images cannot.

Figures 8.5 and 8.6 show imagery we have created, alongside our AltNarratives. When you hover over the imagery, it is not always possible to read the entire script in the yellow pop-up box in *Adobe Acrobat* (Fig. 8.7), although *Microsoft Word* supports this feature if you hover high enough on the screen over the image.

For example, see Fig. 8.5, the AltNarrative:

> A vivid blue eye at the centre of a sunburst reflects a simple oblong in the white pinpoint within the pupil. We zoom in, the white pinpoint grows, and we see it contains images of a textual research paper, pages piled up. The same pages overlap the right-hand corner of the eye, no words are visible, but a bottle of ink has been spilled across the edge of them and a paintbrush lies discarded atop the abstract. The spilled ink is not a single colour, but contains panels of a comic, in it we see people working, interacting. Surrounding the pages and the larger eye is a landscape, lush and green, a small house and fir trees are in the distance.

Fig. 8.5 Introductory image from our AltNarrative paper (Lewis et al., 2022) designed to be viewed without words except for the AltNarrative. Fineliner pen and marker on paper. Miriam Sturdee, 2022

8.4 Additional Techniques for Accessible Sketches 151

Fig. 8.6 Vignette—a day in the life… *Procreate* App on *Apple iPad Pro* using *Apple Pencil*. Makayla Lewis, 2023

Above the ink spillage a girl sits and reads a book, next to a fir tree in the foreground. A cursor sits in the sky, as it pauses the alt text has appeared, but instead of containing text it is the image of an ear and a laptop which is communicating via the screen reader. In the top right corner of the image the eye is repeated again, the red and yellow rays of the sun spanning into the corner of the page. This paper is a celebration of both visual thinking and comics, but also of the stories behind the images, the stories that everyone can hear. We are just beginning this journey however, and so invite your opinions, suggestions and comments to make it even richer. We look at stories and how our images might appear to those who do not have the "ease" of absorbing imagery easily, we think on narratives—AltNarratives—that have been created to make this work enjoyable for as many people as

Fig. 8.7 Page from our AltNarrative paper (Lewis et al. 2022) with the AltText hover-over in *Adobe Acrobat*. Screenshot of Miriam's comic on page 2.

possible, we explore the history of alt text in comics, the problems inherent in asking AI to generate meaningful descriptions, and compare the human and computer in their interpretations of visual imagery and sketches. Many of us champion the use of imagery in research of the pictorial format, but now we need to take this to the next level, let us also champion the use of storytelling to support those images. But the story is just beginning. Let's go to the next page.

For example (see Fig. 8.6), the AltNarrative:

"A vignette of Janet in a variety of locations looking indifferent whilst at and interacting with her phone. Top right: Janet is sitting in a waiting room. It appears to be somewhere official, but the site is unknown; they pause from looking at their phone, looking to the right. It is unclear what they are looking at, but they are checking something. Next, Janet is standing, maybe walking, with a heavy shopping bag on their right arm; whilst holding their phone, they appear to be zooming in on something. They are wearing a smartwatch, indicating the source of smartphone interaction. This is followed by Janet lying in bed looking at their phone; they appear to be reading something. The time of day is unclear, but they appear comfortable; thus, it is likely late evening. The bottom of the image is Janet with a heavy backpack, standing on an empty train platform; the platform is recognisable as they stand on it regularly when commuting to work. In the distance, it appears dark; thus, it is most likely an evening in winter. Finally, the top right is Janet holding theirs, currently visiting the PCS website and completing form fields. Surrounding them are various apps they use regularly, i.e. *Outlook*, *Word*, *X*, *Miro*, *Notes*, *Adobe* and *Procreate*".

Once you feel you have mastered AltText, why not use AltNarrative on some of your more detailed creations from Chap. 6? This alternative accessibility practice is in its infancy, but with more people exploring its potential, we can perfect it together.

8.4.3 Colour Contrast

Your use of colour in your sketches requires consideration; this is referred to as colour contrast or colour difference, i.e. some sketchers and viewers with low or impaired vision or colour blindness are unable to perceive brightness, hue, or saturation between two colours. As a result, considering the colour you use in your sketches is essential; when including people, icons, text, connectors, and separators, you should ensure the line art, background, and surrounding characters, icons, and objects are distinguishable.

The higher your colour contrast, the more accessible your sketches will be. *W3C WCAG* guidelines identify a "contrast ratio of at least 4.5:1" (W17). To meet this requirement, we recommend:

- **Brightness**—the difference in brightness between colours, we suggest that your line art and lettering should have high contrast with background.
- **Saturation**—the intensity of colour, we suggest avoiding positioning highly saturated colours next to other highly saturated colours.
- **Hue**—the difference in colours, we suggest avoiding putting colours with similar hues together; aim for a high hue contrast (Fig. 8.8).

To help you, we recommend identifying the hex code for the colours you intend to use; most digital software and apps will provide you with this information, and for those of you using analogue pens and markers, brands will often provide hex codes for their products, which can be found on their websites. If you cannot find this information, you can swatch your colours on white paper, and then use the *Adobe Capture* App (or a similar app) to identify the hex code. Once identified, input them into a colour contrast checkers, e.g. *Adobe Color Accessibility* tool (W18), and adjust your selections accordingly.

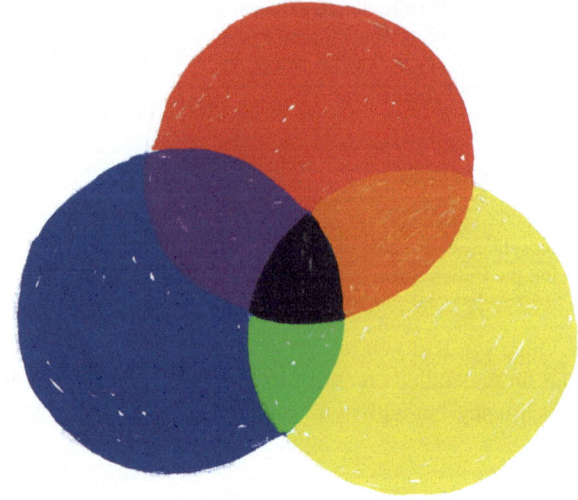

Fig. 8.8 Reminder of colour palette to support colour selection (see Chap. 4). *Procreate* App on *Apple iPad Pro* using *Apple Pencil*. Makayla Lewis

Fig. 8.9 Example of unclear, thin weighted, and italic lettering. *Procreate* App on *Apple iPad Pro* using *Apple Pencil*. Makayla Lewis

Finally, equally important, do not use colour to convey information; if you must do this, provide annotations that textually explain the visualised information.

8.4.4 Typography

Ensuring that your lettering (often referred to as typography or text) is clear, well-weighted and uses accessible style is essential; your sketches lettering and annotations should be legible to ensure viewers can read your work. Thus, it is recommended not to use unclear, thinly weighted, and italic or calligraphy (e.g. Fig. 8.9) but well-spaced and weighted lettering (e.g. Fig. 8.10). Makayla often uses uppercase letters, whereas Miriam uses well-spaced lowercase letters as long as it is clear and legible it is up to you. Finally, do not include icons in words or sentences, as this can confuse many people with or without a disability, impairment or condition.

Fig. 8.10 Example of clear, appropriately weighted, and non-italic lettering. *Procreate* App on *Apple iPad Pro* using *Apple Pencil*. Makayla Lewis

8.4.5 Visual Style and Consistency

Ensuring that your sketches and annotations are consistent within and across your sketches is essential; being consistent in your sketch style (sometimes referred to as design language) can help maintain a unified visual language across your sketches, thus better supporting communication, clarity, efficiency, professionalism, and of course accessibility of your sketches. We recommend following Rob Dimeo's recommendations for language and consistency (Dimeo, 2021):

- **Contrast**—people notice differences—unique elements should be easily distinguishable.
- **Repetition**—people like cohesion—establish a style and use it on similar elements.
- **Alignment**—people like connection—visually connect elements using appropriate alignment, e.g. along paths, with arrows, or speech bubbles.
- **Proximity**—grouping related elements aids comprehension, e.g. placing related elements close to each other, e.g. containers, separators, and clusters of info.

8.4.6 Digitalising

When converting your analogue sketches to digital or creating digital sketches using software or apps, we recommend scanning analogue sketches at 300 dpi or more or setting your digital canvas to 300 dpi or more (see Chap. 14.6 Photographing and Scanning your Sketches). Doing this will ensure that your sketches are clear and can be zoomed in without blur (pixelation). This will help sketchers and viewers who wear glasses or use digital magnifiers to better interact with your digital sketches. Miriam and I wish we had considered this ten years ago. Thus, we apologise for some of our earlier work (presented in this book and online that is not high resolution).

8.5 Hands-On Activities

Activity 8.1: Checking Your Colour Contrast (Individual)
Learning objective—To evaluate your choice of colours
Time—15 minutes
Materials—Analogue or digital pens and markers, hex codes, and *Adobe Colour Accessibility* Test
Procedure:

- Pick your favourite fineliner pen and at least two markers of different colours.
- Identify the hex codes for your pen and marker colours.
  - Most digital software and apps will provide you with this information, and for those of you using analogue pens and markers, brands will offer hex codes for their products, which can be found on their websites.
  - If you cannot find this information for your analogue colours, swatch your colours on white paper, and then use the *Adobe Capture* app to identify the hex code.
- Insert your pen hex code and, in turn, one of your marker hex codes, and look at the results; if favourable, note that these colours can be used together; if not, try a different marker.

Activity 8.2: Create AltText (Individual)
Learning objective—To become competent in describing your images for those using a screen reader
Time—Up to 5 minutes per sketch
Materials—You will need to reference your images from previous chapters. Choose your favourites, or ones which you might like to share online
Procedure:

- Return to one or more of your previous sketches.
- Look at the guidance presented in this chapter and create AltText for your sketch.
- If you use social media, utilise the tool to add the AltText, and release it into the wild!
- If you prefer, start a new document in *Microsoft Word* or *Adobe Acrobat* (PDF), and practise adding AltText.
- Going forward, make it a part of your standard practice to always write AltText for shareable documents.

Below (Fig. 8.11) is an example from Chap. 3 that we have revisited, on developing icons and building sketched detail. The AltText that could accompany this could be:

A digital black and white sketch showing the development of an icon sketch of a passport. There are three iterations, with the progression shown with red arrows between each version. Version one shows the outline of a passport with a few lines to show pages underneath and text on the front cover. Version two adds a squiggle to represent the coat of arms or symbol of the country the passport belongs to. Version three adds final details such as a biometric passport symbol and extra lines above the country symbol.

Fig. 8.11 Passport icon detail development. *Procreate* App on *Apple iPad Pro* using *Apple Pencil*. Miriam Sturdee, from Chap. 3, copy of Fig. 3.3

Activity 8.3: Create AltNarrative (Pair or Group)
Learning objective—Exploring the potential of AltNarrative
Time—Up to 10 minutes writing per detailed sketch, 10–15 minutes sketching, 5 minutes discussion
Materials—You will each need to reference your visual narrative examples from Chap. 6. Choose your favourite or most complex visual narrative!
Procedure:

- Each person in the pair or group should use their visual narrative as a resource—do not share your chosen image with your partner or group!
- Looking at the examples in this chapter, create AltNarrative for your visual narrative, either type and print or write out in clear, legible text.
- Swap AltNarrative with your partner or another group member.
- Each person should now have a detailed AltNarrative to read.
- Without referencing the other person's image, sketch out the visual narrative using only the AltNarrative.
- When finished, compare and contrast the visual narrative to the original, and consider what has changed. Did you add anything extra? Was the rich description helpful?
- Discuss as a group what aspects of AltNarrative were the most successful, and create a best practice list which can be shared with a wider group or class.

References

Books, Papers, and Articles

Dimeo, R. (2021). *Sketchnoting science: How to make sketchnotes from technical content.* nvl-pubs.nist.gov/nistpubs/SpecialPublications/NIST.SP.1265.pdf

Lewis, M., Sturdee, M., Miers, J., Davis, J. U., & Hoang, T. (2022, April). Exploring AltNarrative in HCI imagery and comics. In *CHI conference on human factors in computing systems extended abstracts* (pp. 1–13).

Websites

W1 World Wide Web Consortium is the international standards organization for the World Wide Web - WCAG standard – www.w3.org/press-releases/1997/ipo-announce/
W2 United Nations definition of disability – www.un.org/development/desa/disabilities/convention-on-the-rights-of-persons-with-disabilities/article-2-definitions.html
W3 UK government definition of accessibility – www.accessibility.blog.gov.uk/2016/05/16/what-we-mean-when-we-talk-about-accessibility-2/
W4 Twitter (X) support page for adding image descriptions – www.help.twitter.com/en/using-x/add-image-descriptions
W5 Instagram support page for adding image descriptions – www.help.instagram.com/503708446705527
W6 TikTok support page for autocaptions – www.newsroom.tiktok.com/en-us/introducing-auto-captions
W7 Youtube support page for auto captions – www.support.google.com/youtube/answer/2734796?hl=en-GB
W8 World Accessibility Day comments on accessibility www.accessibility.day
W9 UK government definition of disability – www.gov.uk/definition-of-disability-under-equality-act-2010
W10 Olympic museum pictorgrams from the 1964 Olympics – www.olympic-museum.de/pictograms/olympic-games-pictograms-1964.php
W11 UK government accessibility posters for designers – www.ukhomeoffice.github.io/accessibility-posters/
W12 World Wide Web Consortium - Web Accessibility Perspectives (Compilation of 10 Topics/Videos) – www.youtube.com/watch?v=3f31oufqFSM
W13 Web AIM, non-profit organisation, comments on Alt Text creation – www.webaim.org/techniques/alttext/
W14 Adobe support page for adding image descriptions – www.community.adobe.com/t5/acrobat-discussions/adding-alt-text-to-figures-adobe-acrobat-pro-dc/td-p/8585014
W15 Microsoft Office support page for adding image descriptions – www.support.microsoft.com/en-au/office/add-alternative-text-to-a-shape-picture-chart-smartart-graphic-or-other-object-44989b2a-903c-4d9a-b742-6a75b451c669
W16 World Wide Web Consortium is the international standards organisation for the World Wide Web - WCAG 2.0 www.w3.org/TR/WCAG20-TECHS/PDF1.html
W17 World Wide Web Consortium is the international standards organisation for the World Wide Web - WCAG 2.20 colour contrast – www.w3.org/WAI/WCAG21/Understanding/contrast-minimum.html#:~:text=The%20visual%20presentation%20of%20text,Incidental
W18 Adobe colour contrast tool – www.color.adobe.com/create/color-contrast-analyzer

Further Reading, Viewing, Listening!

Books

Henry, S. L. (2007). *Just ask: Integrating accessibility throughout design*. Lulu.com.
Horton, S. and Quesenbery, W., 2014. A web for everyone: Designing accessible user experiences. .
Kalbag, L. (2017). *Accessibility for everyone* (pp. 8–117). A Book Apart.

References

Podcast

How to Conduct Inclusive UX Research: www.careerfoundry.com/en/blog/ux-design/inclusive-ux-research/

Designing for Users With Low Vision by Michele A. Williams and Synge Tyson #id24 2020 www.youtube.com/watch?v=EMgqHEu5MK4&list=PLn7dsvRdQEfGkK9xxk54XdKTLk7zf_Qwp&index=4

Let's Inclusify / Rachel Rodney #id24 2020 www.youtube.com/watch?v=yBir_QF_sk8&list=PLn7dsvRdQEfGkK9xxk54XdKTLk7zf_Qwp&index=5

Microsoft Research 'Disability Types' www.youtube.com/watch?v=QJgca3Sb8iQ

Websites

Accessibility Stories and Interviews by Deque: www.deque.com/empathy-lab-online/accessibility-stories/

How making services accessible benefits all users by Alistair Duggin: www.youtube.com/watch?v=myczHDuBHuY&feature=youtu.be

A international community project to make digital accessibility easier www.a11yproject.com/

Suite of integrated accessibility toos for product design www.getstark.co/

Accessibility WAVE tool to evaluate the accessibility of websites www.chrome.google.com/webstore/detail/wave-evaluation-tool/jbbplnpkjmmeebjpijfedlgcdilocofh/related

RNIB UK accessibility guidelines www.rnib.org.uk/accessibility-guidelines-alt-text-what-you-need-know

Chapter 9
Digital Sketching Techniques

9.1 Getting Started

Digital sketching still feels relatively new to many people, despite its invention with Ivan Sutherland's "SketchPad" in 1964 (Sutherland, 1963). The technology has come on leaps and bounds in the past 15 years. Now, you might as well see someone taking out a tablet to sketch on as a piece of paper or sketchbook. Many artists made the leap as digital sharing became commonplace, and sketching directly into digital removed the need for cumbersome image scanners. However, there is an immediacy and a tangible nature to the pen and paper sketch that has never been fully realised in digital form, despite efforts to add "paperlike" screen protectors or rougher surfaces. We could debate that digital sketching devices should try instead to embrace their nature as un-paperlike, but many of us crave the dual experience.

Digital sketching is not a cheap option, which is one of the many reasons why we advocate pen and paper as a starting point. The cost is a barrier to access, with the best machines and programs adding further cost to the most basic starting point. Like most digital technology, these devices also have incremental updates, year to year, so to recommend a definitive purchase in this book would be irresponsible.

By the end of this chapter, you should be able to:

1. Consider the pros and cons of digital versus analogue sketching.
2. Choose an appropriate device and approach that works for your personal style.
3. Customise your virtual experience to suit a variety of situations.

9.2 What's New, What's Different

Unlike analogue sketching, digital sketching has many unique advantages:

- **Versatility and efficiency:** Digital sketching apps provide access to a variety of "paper" (also known as canvas' in the digital world), brushes (pens, markers, etc.), and effects (watercolour, stamps, etc.) that are realistic but also abstract that can be personalised and switched between quickly and easily.
- **Unlimited undo:** Unlike analogue sketching, digital sketching apps allow the sketcher to delete and correct their sketches without wasting materials, as per the meme, "no one will know", thus supporting worry-free experimentation/practice.
- **Infinite canvas:** Digital canvas dimensions can be modified as required by the sketcher; this unlimited canvas means you can sketch without restriction. You no longer need to search for Sellotape or glue sticks to attach two or more pages together because the sketch is larger than you anticipated.
- **Sharing:** Digital sketches are quick and easy to share between devices, online social media and productivity platforms, software, and with users and collaborators without digitalising.
- **Time-lapse recordings:** Makayla loves this feature (e.g. Sturdee et al., 2021); the ability to review your sketch process is lovely, as it will allow you to streamline, improve, and share your workflow/process—supports self-reflection and growth.
- **Storage and organisation:** We recommend storing sketches in one place, preferably in the cloud, e.g. *OneDrive*, *DropBox*, *Google Drive*, etc. Do not forget to give your files logical names; otherwise, it's like having tons of sketches scattered across a room (loose and in sketchbooks),—a needle in a sketch "haystack" so to speak. You will thank yourself later.

Whilst digital sketching offers many benefits, it's important to note that some sketchers prefer the unique qualities of analogue sketching; there is nothing wrong with that. The choice between digital and analogue sketching depends on your preferences, goals, and, in some cases, the specific project.

9.3 Miriam's Story

In terms of digital sketching, I am a beginner. I have owned various devices in the past 9 years… and I am STILL a beginner. I started with a *Cintiq Companion 2*, heralded as the nearest thing to real paper you could get. I even made a comic with it, experimenting by using *Photoshop* to sketch directly in for the first time (Sturdee et al., 2016) (Fig. 9.1). Then I went straight back to pen and paper for my PhD work and got heavily involved with Sketching in HCI with Makayla, where we celebrated the analogue AND the digital. The Cintiq had a battery issue that was not covered

9.3 Miriam's Story

Fig. 9.1 Miriam's first digital image—a real chance to experiment with digital sketching! from (Sturdee et al., 2016). *Cintiq Companion 2*

by the warranty, and although it was used for some digital sketchnotes for ACM Interactive Surfaces and Spaces (Fig. 9.2), it was no longer portable, nor practical.

You can see the difference in my sketchnotes between Chap. 6 and here. The style is very rough, and I find I was much less confident committing to a line than if I was using pen and paper. I felt there was a disconnect between my ability and the page. In Fig. 9.3 however, I simply sketched on paper and photoscanned the images into an application, before applying fill and brush effects in *Photoshop* later and adding the background (also *Photoshop*).

Our Sketching in HCI courses have taught me a lot about the benefits of digital sketching, especially when communicating to diverse audiences in-person and remotely. The set up we usually had for the course when it was entirely in person (pre-pandemic) was that Makayla would connect digitally via *iPad* to the projector screen and I would be sketching along on flip charts and whiteboards as the venue allowed. This dual process shows both our approaches to sketching at the same time and enriches the experience.

When we flipped to remote sketching classes, I began to use a *Microsoft Surface Go 3* to interact directly with a digital whiteboard or set up my phone on a flexible arm to record real-time analogue sketching. However, the *Surface Go 3* lacked the processing power for more memory-intensive programs, and whilst the Microsoft stylus is a nice piece of kit, it became a tool of necessity rather than a joyful

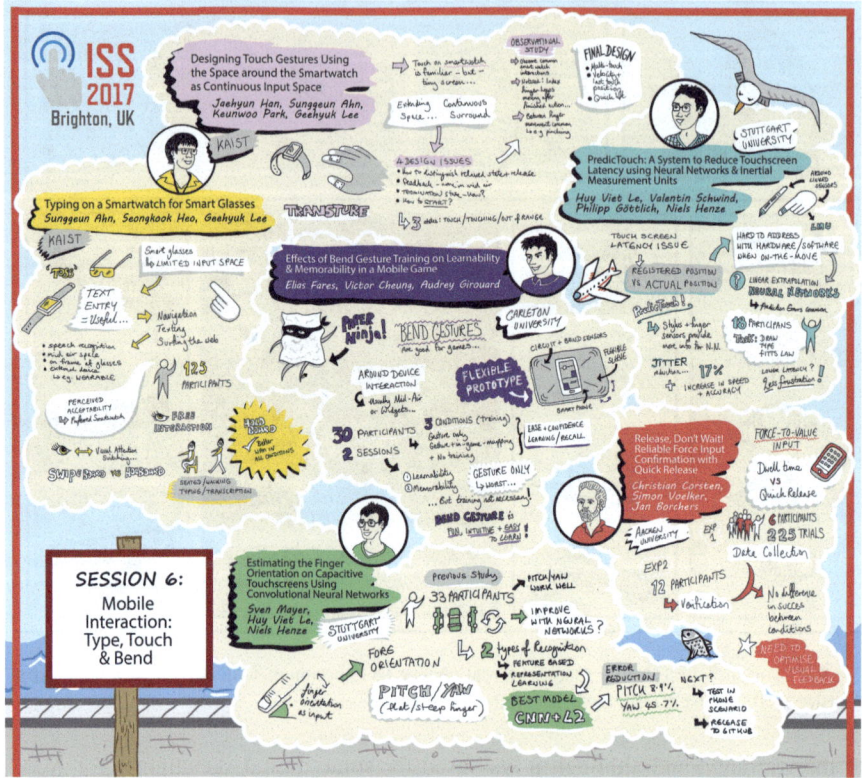

Fig. 9.2 Digital sketchnotes for ACM ISS 2017, *Cintiq Companion*. Sketches are on a preprepared canvas; this also shows examples of using white on colour and taking advantage of layering. *Photoshop* on *Microsoft Surface Pro* using *Microsoft Surface Pen*. Miriam Sturdee, 2017

experience. However, it did allow me to sketch several fun images based around an interactive "egg world" hosted in *Gathertown* (W1) (Lindley et al. 2021) (Fig. 9.4).

As I write now, I have been given an *iPad Pro* with *Apple Pencil* by my department, and I have downloaded (a paid-for app) *Procreate* App. I find the change in interface difficult but am learning how the brushes work, how you can edit the pressure and line, and how to work with layers and colour. Some images in this book may even be sketched digitally as I get to grips with this third tablet tool.

For me, the sketching experience is all about paper though. The digital approach is useful, but the device is a tool rather than something I will reach for automatically. If I need a digital, layered image or to overlay some different colours which would simply be lost if I tried markers, I can utilise this tool. I love the feel of a pen and the ink on paper; I love switching coloured markers from the pile on my desk; I love the immediacy, unpredictability, and organic nature of watercolour or ink. So much of our lives are already in the digital domain; why not keep something back?

9.3 Miriam's Story

Fig. 9.3 Hybrid approach to sketching digitally. Loose sketches were taken in a sketchbook with black fineliner and then imported into the Moleskine iPhone application (now discontinued) and rough "fill" effect added (black) and white highlighting added manually. *Photoshop* on *Microsoft Surface Pro* using *Microsoft Surface Pen*. Miriam Sturdee, 2017

Fig. 9.4 Digital sketch of a "room" in the egg part of a "not paper" hosted in Gather Town (Lindley et al., 2021). *Photoshop* on *Microsoft Surface Pro* using *Microsoft Surface Pen*. Miriam Sturdee, 2020

9.3.1 Miriam's Practical Application Tips (for the Digitally Shy Sketcher)

- Fear of the blank screen is the same as fear of the blank page! Try making a new file and trying out some different effects, brushes, and colours.
- Revisit your inner child from Chap. 2 and scribble, lines, shapes, and spirals.
- Don't be afraid to mix things up—combining scanned hand-sketched imagery and digital layering and annotation can be a great way to explore and express concepts.
- If you find yourself sketching in a different style than usual, don't try to force your analogue style; embrace the new way of working—you can have a style for each approach!

Miriam's Tools for Digital Sketching

Tools:

- *Apple iPad Pro 11″* and *Apple Pencil* (at home)—the *Surface Go 3* has been retired!

Software:

- *Procreate* on *iOS* (one-time payment).
- *Adobe CC Photoshop* (subscription).

Support devices:

- *AirDrop* to move files between devices and software (*Apple* only).
- *Paperlike* screen protector for *iPad*—offers realistic friction.

I hope to discover more items as I continue my digital sketching odyssey! Further useful tools are listed in Makayla's section below.

9.4 Makayla's Story

Compared to Miriam, I have become a 70/30 sketcher; most of my sketches are done digitally. This is because it is convenient and quick when on the go, e.g. during teaching, field research, or in a busy coffee shop. Carrying my *iPad Mini* and *Apple Pencil* gives me a variety of "paper", "pens", and "markers" without having to take an assortment of paper, pens, and markers. My backpack shoulder is often pleased by this decision. When at my workstations, I prefer large devices as they support better posture and allow me to get an overview of my sketches. I use my *iPad Pro* at home, and I use *Wacom Cintiq Pro* in the university office. Don't get me wrong; I love analogue materials; I often return that to analogue when doing observational sketching, narrative drafting, sketch notes, and just fun/practice. Digital sketching became part of my life in 2013; I struggled to try different software and devices since, at first, it was frustrating. I knew what I wanted but getting it onto a screen was impossible; I was never happy. To build my skills, I opted to take the hybrid sketch approach, i.e. sketch using analogue materials, scan work, and then editing digitally. This allowed me to create visual representations that I saw in my mind on paper and then improve the outputs digitally, e.g. adding backgrounds and editing or deleting elements. This approach allowed me to build my digital sketching skills whilst also being happy with the work I was producing. For examples of my digital sketches, see Figs. 9.5, 9.6, 9.7, 9.8 and 9.9.

In 2020, the first COVID-19 lockdown changed my mindset; I had the time to step away from the hybrid approach, separating the two sketching forms. For those first 6 weeks, I only digitally sketched, took classes on *Skillshare*, downloaded a bunch of "how-to books" on my *iPad*, began collecting *Character Design Quarterly* (a magazine I still collect to this day, thank you 3DTotal Publishing and the artist that have shared their digital work, process, and tips). I believe digital sketching allows for a gentle learning curve; the numerous tutorials and resources and easy access support (how-tos) make exploring sketching and developing skills easier, e.g. Fig. 9.10. Embracing digital during 2020 kept my mind and hands active when not working, which helped create a more sustainable work-life balance when lines blurred for many of us.

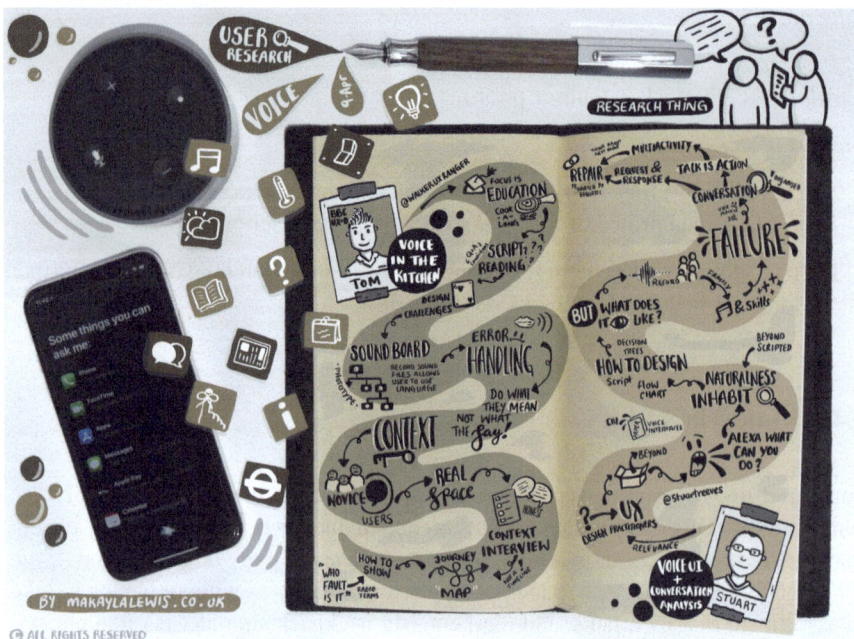

Fig. 9.5 Sketchnotes from Research.Think meetup "User Research and Voice". *Procreate* App on *Apple iPad Pro* using *Apple Pencil*. Makayla Lewis, 2020

9.4.1 Makayla's Practical Application Tips for Digital Sketching

To help you improve your digital sketching skills, I recommend the following:

- **Hardware:** A tablet with a pen (e.g. *Apple iPad* and *Pencil* or *Microsoft Surface* and *Microsoft Surface Pen*) is much easier to use than a graphic tablet (e.g. *Wacom Intuos*). It can enhance your digital sketching experience by offering more precision and control. I do not recommend using a mouse or trackpad. I have tried and commend those who can; my sketches were a legible mess.
- **Software:** There are various digital sketch apps and software options available, but there is no right one to use; start with the free ones, and then explore further, and choose the one that suits your style, needs, preferences, and budget.
- **Customise digital workspace:** Familiarise yourself with your chosen app or software user interface using online tutorials, resources, short courses, etc. (see **Further Reading** for examples). I recommend customising your digital workspace to your liking is essential as it will make it easier to use, and thus create sketches, e.g. setting up shortcut keys, pinned brushes, and colour plates for quick and easier access. When digitally sketching, you will often stick to the same "brushes" (pen, pencil, marker, etc.), canvas size, and colours, thus making selecting them quicker, which in turn save time, thus streamlining your process.

9.4 Makayla's Story

Fig. 9.6 Sketchnotes from Research.Think meetup "User Research and Kids". *Procreate* App on *Apple iPad Pro* using *Apple Pencil*. Makayla Lewis, 2020

- **Simple digital tools:** Use the stock "brushes" (pen, pencil, marker, etc.) provided by the app or software provider, round brush (e.g. *Procreate* App Studio Pen for line art, sketch brush (6B pencil) for drafting, and round eraser to erase (or you can use the two-finger tap in *Procreate*) is often a good starting point. Once confident with these simple tools, modify them and try other brushes and effects.
- **Layers and workflow:** One of the most important aspects (learned the hard way) is the importance of using layers when digitally sketching. It took me a while to realise that using different layers for different elements of your sketch (background, foreground, characters, and colours) makes the process easier to delete and edit individual components without affecting the entire composition. I cannot stress this enough: USE LAYERS! To help you establish a workflow that works for you; you will do this through practice. For example, I order and name my layers before I start:

  Layer 7—Text/Lettering.
  Layer 6—Line Art (Foreground)—sometimes, I will mask this layer, but it is not necessary for simple sketches.

Fig. 9.7 ACM SIGCHI workshop "The future of computing and food". *Photoshop* on *Microsoft Surface Pro* using *Microsoft Surface Pen*. Makayla Lewis, 2019

Fig. 9.8 Close up of Makayla sketching a comic for an ACM SIGCHI 2022 paper (Lewis et al., 2022). *Procreate* App on *Apple iPad Pro* using *Apple Pencil*, Makayla Lewis 2021

9.4 Makayla's Story

Fig. 9.9 Sketchnotes book summary "How to draw anything". *Procreate* App on *Apple iPad Pro* using *Apple Pencil*. Makayla Lewis, 2019

> Layer 5—Colour (Foreground).
> Layer 4—Line Art (Background)—sometimes, I will mask this layer, but it is not necessary for simple sketches.
> Layer 3—Colour (Background).
> Layer 2—Blue Pencil Draft.
> Layer 1—White Background.

- **Seek feedback and confidence:** I think Miriam would agree to ask for help (and seek constructive feedback) from fellow digital sketchers; it can help you identify areas for improvement and growth. Miriam always asks me: *how do you do that? Is this the most effective way to do this?* Do I mind? Nope; keep the questions coming Miriam. Remember to enjoy your sketch process, and do not be too hard on yourself (Miriam, did you read that?).
- **Save and backup:** It is rare, but digital files can become corrupted (or accidentally deleted—I have done this a couple of times); thus, save your work regularly. I highly recommend using cloud-based storage and saving as a PNG when complete.

Fig. 9.10 A sketch exploring of the perfect Christmas Mince Pie. *Procreate* App on *Apple iPad Pro* using *Apple Pencil*. Makayla Lewis, 2020

Makayla's Tools for Digital Sketching

Tools:

- *Apple iPad Mini* and *Apple Pencil* (on the go).
- *Apple iPad Pro 11″* and *Apple Pencil* (at home).
- *Wacom Cintiq Pro 22″* and *Wacom* Pen (at university office).

Software:

- *Procreate* on *iOS* (one-time payment).
- *Adobe Creative Cloud Photoshop* (subscription).

Support devices:

- *BRÄDA* Laptop support for *iPad Pro 11″* for better posture.
- *Apple Smart Folio* for *iPad Mini* for better posture but not as good as *BRÄDA* Laptop support.
- *Ergon Arm* for *Wacom Cintiq* for better posture.
- *AirDrop* to move files between devices and software.
- *OneDrive* to store and organise files.
- *Procreate* a mini keyboard to access shortcuts faster.
- Paperlike screen protector for *iPad* tablets.

As digital sketching tools innovate, I look forward to experimenting with new and unique possibilities.

9.5 Hands-On Activities

Activity 9.1: Analogue vs. Digital (Individual)

Learning objective—Build confidence in creating simple sketches using digital tools

Time—Up to 2 minute per icon in the first phase, up to 2 minute per icon in the second phase, up to 2 minute per icon in the third phase

Materials—Digital sketching device (can be borrowed) and stylus, five of your icon sketches from Chap. 3

Procedure:

- Pick five of your favourite icons from Chap. 3.
- Using a sketching program of your choice, recreate each icon, WITHOUT using the undo function. You may, however, use zoom and rotate if needed.
- When you have finished all five, compare them to the original; ask yourself, how do the lines differ, was it easier or harder to add details? Which do you prefer?
- Now redraw each icon with a different digital "brush" (phase 2); do the different weights or effects lend themselves to some icons more than others? What happens if you use multiple brushes on the same icon?
- Finally, copy and paste your favourite digital version of the icon, and experiment with colouring it in—you may use the layer function if the program allows, and yes, you can now use "undo"!

For added fun, you can also try to draw the same items using your finger in place of the stylus… does the tactile feedback help or hinder?

Activity 9.2: Digital Reference Library (Individual)

Learning objective—Create a go-to image of brushes and effects, with reference

Time—30 minute

Materials—Digital sketching device (can be borrowed) and stylus

Procedure:

- Choose a digital sketching program that you have regular access to (where possible).
- Create a new file; if applicable, use a grid guideline to help keep your sketches neat and tidy.
- For each "brush" effect (not just brush—this term includes pen, pencil, and so forth), create a swatch giving examples at different weights and one colour example.
- On a different layer, or file, have a "scribble board" much like we created in Chap. 2. This will help you get a feel for each type of mark maker.
- Depending on the program, this could take some time, so initially try to choose brushes you think will be the most practical in your day-to-day practice.
- You can make this a work in progress and revisit the file as you continue on your digital journey.
- If you like to have PHYSICAL visual prompts, print out a colour copy in high resolution, and pin it to a noticeboard, or stick to the wall.

Remember to learn how to save your preferences if the program allows! This will save you time when you have to quickly create a sketch for a particular output.

Activity 9.3: AltText/AI Telephone Game (Group)

Learning objective—Become confident at quick-fire digital sketching
Time—10 minute (30 s per sketch)
Materials—Digital sketching device (can be borrowed) and stylus, access to email or chat that accepts images and text, a program that generates auto-AltText (although an AI program may be used if desired)
Procedure:

- Participant 1 creates a VERY quick line sketch of a simple scene on their digital device (e.g. a dog eating an apple) and places it into a program such as *PowerPoint* which makes AltText suggestions. They then send this suggestion to participant 2.
- Participant 2 sketches their interpretation of the AltText and generates a new AltText description.
- Participant 3 repeats the process, and so on until each group member has had a turn.
- All sketches are revealed (hopefully to some hilarity!).

Large language models have changed the way we interact digitally—and with sketching. In fact, human-AI "collaborations" are becoming more commonplace. You can make this game even more silly and fun by using an agent instead of the basic AltText generators. We talk more about AI and the future of sketching in Chap. 13.

References

Books and Papers

Lindley, J., Sturdee, M., Philip Green, D., & Alter, H. (2021, May). This is Not a Paper: Applying a Design Research lens to video conferencing, publication formats, eggs… and other things. In *Extended abstracts of the 2021 CHI conference on human factors in computing systems* (pp. 1–6).

Lewis, M., Sturdee, M., Miers, J., Davis, J. U., & Hoang, T. (2022, April). Exploring AltNarrative in HCI imagery and comics. In *CHI conference on human factors in computing systems extended abstracts* (pp. 1–13).

Sturdee, M., Coulton, P., Lindley, J. G., Stead, M., Ali, H., & Hudson-Smith, A. (2016, May). Design fiction: How to build a Voight-Kampff machine. In *Proceedings of the 2016 CHI conference extended abstracts on human factors in computing systems* (pp. 375–386).

Sturdee, M., Lewis, M., Strohmayer, A., Spiel, K., Koulidou, N., Alaoui, S. F., & Urban Davis, J. (2021, June). A plurality of practices: Artistic narratives in HCI research. In *Creativity and cognition* (pp. 1–14).

Sutherland, I. E. (1963, May). Sketchpad: A man-machine graphical communication system. In *Proceedings of the May 21–23, 1963, spring joint computer conference* (pp. 329–346).

Websites

W1 Gathertown, an online space to meet and host events – www.gathertown.com
W2 For digital sketching tutorials – www.youtube.com/
W3 For digital sketching tutorials – www.skillshare.com
W4 to find digital sketching tutorials and process – www.patreon.com

Further Reading

Kelkar, U. (2020). *The urban sketching handbook drawing with a tablet: Easy techniques for mastering digital drawing on location* (Urban sketching handbooks). Quarry Books.

Publishing 3dtotal. (2017). *Character design quarterly*. 3DTotal Publishing (issues 1 onwards).

Publishing 3dtotal. (2023). *Beginner's guide to procreate: Characters: How to create characters on an iPad*. 3DTotal Publishing.

Ulichney, M., Grünewald, S., Stokart, A., & Nassour, S., (2020). *Beginner's guide to digital painting in procreate: How to create art on an IPad*. 3dtotal Publishing.

Chapter 10
Remote Sketching

10.1 Introduction

The world of meetings, events, seminars, conferences, and lectures were turned on their head during (and a result) of the COVID-19 pandemic and subsequent lockdowns. Rather than nipping along to your colleague's office to ask a quick question, chat functions have become the norm, with organisations using digital products and services such as *Microsoft Teams* (W1), *Google Drive* (W2), or advanced hosting software such as *Zoom* (W3) and *Google Meet* (W4). Online meetings spread from a rare event during the week to full days in front of the screen, which also meant online teaching and learning. So how did this impact the delivery of creative, hands-on courses, such as sketching? And given the new "hybrid" world we live in, how can we utilise these changes to maintain the impact of sketching in HCI?

Both as educators and learners, remote engagement with sketching enables us to continue our creative practice, whatever the climate. As educators, it can help us to broaden audiences, reaching those who live on the opposite side of the world; as learners or participants, it can allow courses to fit around personal lives, family, and work. Although there are barriers to providing materials, sketching in HCI can be a good pastime, as some of our most engaged learners have used a simple rollerball and waste paper.

This section provides a helpful list of tools, techniques, and materials you might need on either side of the course delivery, alongside four case studies of our own experiences delivering Sketching in HCI education as a stand-alone or as part of other, broader courses such as UX. We then explore activities that work well in the remote context and are still engaging, creative, and—above all—still fun.

10.2 Tools and Recommendations

The simple fact of remote sketching is that all involved need some form of Internet-connected hardware and software. Although the list below provides an "ideal" list, engaging with just a laptop and physical materials—even without a webcam—is more than possible, as individuals can use online whiteboards to share sketches. However, most modern hardware does contain an in-built camera, which helps with face-to-virtual-face engagement. Miriam who prefers analogue—pen and paper—sketching discovered, you can purchase "arms" to hold your mobile device over the table and live stream sketching (see Fig. 10.1). Tools are very much a personal preference, and this list can advise, but you will find your preferred setup—and once you have this in place, why not sketch it (e.g. Figs. 10.1 and 10.2)?

Environment and mindset are also wholly individual, but for those who prefer to organise their space and time, the notes below offer simple guidelines to prepare yourself. Some of you may also find yourselves hosting sessions to showcase your new understanding and practice or share with remote friends or colleagues for fun.

Hardware

- Mac or Windows desktop or laptop.
- External second monitor.
- HD webcam.
- Visualiser or webcam and arm.

Fig. 10.1 Online learning sketching setup in a quiet, well-lit, location: 27" monitor, external microphone, external webcam, external keyboard, external mouse, and *iPad Pro* on a laptop stand. *Procreate* App on *Apple iPad Pro* using *Apple Pencil*. Makayla Lewis, 2021

10.2 Tools and Recommendations

Fig. 10.2 Miriam's pandemic desk setup: *Microsoft Surface Go* and stylus, *Macbook Pro*, 20″ monitor, mobile phone, table clamp for live hand-drawn view, pens, paper, desk lamp. Fineliner pen and marker on paper. Miriam Sturdee, 2021

- Keyboard.
- Mouse.
- Smartphone with camera.
- Broadband.
- Drawing tablet and digital pen, e.g.:
  - *iPad Mini*, *Air*, or *Pro* with *Apple Pencil*.
  - *Microsoft Surface Go* or *Microsoft Surface Book* with *Microsoft Surface Pen*.
  - *Wacom One* or *Cintiq*.

Software

- An online video conference platform, e.g. *Zoom*, *Teams*, *Google Meet*.
- An online collaboration platform that supports drawing with a digital pen, e.g. *Miro* (W5), *Mural* (W6), *FigJam* (W7):
  - Digital drawing pen (black, adjustable weight).
  - Digital highlight pen (grey and pastel colour).
  - Digital eraser.
  - Digital sticky notes.

Environment

- Well-lit space.
- Ergonomic chair and desk setup to ensure a comfortable session.
- Limited distractions.

Facilitator and Participant Mindset

- Be open-minded and creative in thinking about and seeing the world around you.
- Remember, practice makes progress—take responsibility for improving your practice and learn from your mistakes.
- Embrace curiosity and exploration.
- Discourage judgement and criticism.
- Stay engaged by using digital tools like you would any other sketching tool.

Facilitator Skills

- Good communicator, be clear and concise with instruction and demonstration.
- Listen, do not interrupt others.
- Responding to questions in multiple formats (voice and chat).
- Manage your time and stress levels.
- Set your learning goals before you begin.
- Be flexible in your approach to match the needs of your participants.
- Embrace in suitable humour to encourage a fun, relaxed, and created environment.

10.3 Remote Sketching Case Studies

During the pandemic, we were required to pivot their teaching to online; this included sketching sessions, something normally held in person. Due to the development of online whiteboards, this transition was softened, as these programs allowed for synchronous, live on-screen sketching, which learners could follow in real time and add their own creations to the board alongside that of the instructors.

During the pandemic, we had to pivot their teaching to online; this included sketching sessions, generally held in person. Due to the development of online whiteboards, this transition was softened, as these programs allowed for synchronous, live on-screen sketching, which learners could follow in real time and add their creations to the board alongside that of the instructors.

Case Study 10.1: Undergraduate Sketching in HCI Class
This synchronous sketching class was a 1-hour, live, hands-on sketch-a-long. Slides were used to structure the session, but these were not shown to students due to the limitations of the *Microsoft Teams* environment at the time. The main window was set up to screen share a *Microsoft Surface Go*, and the meeting joined via an additional device—a laptop and second screen to view the slides and also maintain the "chat" function to answer questions. The session covered basic sketching visual-vocabulary (e.g. people, actions, places, animals, objects) and moved on to creating narratives

and storyboards, which directly related to the coursework component. Students all used their own materials but were given advance notice on what to have (paper, black pen, coloured pen, as a minimum). Prior to the session, students with accessibility needs were spoken to privately and alternatives and accommodations arranged (e.g. using a particular stylus and tablet, digital variations on sketching, embedding clip art, and line drawing to create narratives—revisit Chap. 8 for more on accessibility in sketching). There were 276 students in the cohort and, 180 students joined and engaged with the synchronous session and although the images were non-deliverables, some students included these sketches within their coursework appendix, as well as evidencing items from their learning in coursework user storyboards, such as the use of highlighting interactions and framing different viewpoints.

To further support the students' sketching practice, a second synchronous task was delivered, based upon our "HCI Improv" activity. During our usual peer-to-peer sessions (see Case Study 10.3), people form teams and work on spontaneous prompts suggested by the full cohort, before ideating, diagramming, and creating storyboards for a novel technology, use case, and user group. They then present these to the room. For the online setup, the students provided the prompts, but the sketching was done live by the teacher, and students could advise on aspects of the technology and use case in real time using *Microsoft Teams* chat. This part of the course was not designed to teach sketching but to demonstrate its value and instil practical knowledge of HCI and design thinking in an engaging way.

Following the online synchronous sketching lecture, several students reached out to state how much they enjoyed the content and that they had not expected to have so much fun in a computer science lecture. The overall course feedback was positive, with HCI achieving an average score of 4.45/5 for the module, based on the standard university metric—this was the highest score the course had ever had, with a lot of feedback about the alternative activities and sketching skills. Although one student mentioned they thought that the "importance of sketching was overstated" compared to the bulk of the lecture and coursework material (!), it was also incorporated heavily into the open-book exams that were taken by students 6 months later, at which point it became clear why so much focus was on practical sketching skills. Several students actively sought out opportunities to work as teaching assistants for the following year, based on their enjoyment of the course and, in particular, sketching skills and their applications.

Case Study 10.2: Embedding Technology into Sketching Education During COVID-19 Pandemic
As a result of the UK government's online learning measures, the traditional in-person module structure and schedule were overhauled. The Makayla's decision-making process was supported by personal observations and experiences alongside learnings shared the wider UX education community during previous lockdowns; these nonacademically published mediums include blogs, social media (e.g. the *X* hashtag #onlineteaching), articles, virtual department exchanges and coffee breaks, and institution teacher training. The conclusion that online module students would experience challenges that would impact their ability to learn to sketch would

include, but were not limited to, passiveness; time management and discipline; learning environment control; isolation, anxiety, and depression; lack of motivation; and reduction in help seeking. To overcome these, the module structure and schedule were planned and delivered to promote technical and social presence; the module teaching team (the teacher supported by a teaching assistant, a previously successful module learner) actively and regularly engaged and encouraged the company. Although not a new concept, initially put forward by Mishra (2020), it was fundamentally and logistically different from the teaching team before module delivery. Similar to Case Study 10.1, the teacher consulted with students who identified as having accessibility needs before the commencement of the module, and the module materials were adjusted accordingly.

The students were divided into 20 groups, 4 to 6 members, using the People feature within Canvas, a course management system that supports online learning and teaching. Each group was given a link to a private sub-channel within the *Microsoft Teams* module. It was a space for private conversations with a specific audience where they were encouraged to collaborate through chat and video, inside and outside class hours. It was believed that providing the learners with a symbolically "hidden" online space allowed Makayla to separate the large cohort into smaller groups to promote a learning environment conducive to confidence and relationship building in a supportive peer environment. This space would encourage exploration and experimentation with sketching. Makayla created a teaching space in a quiet, well-lit location of their home to minimise distractions during online teaching sessions. Sketching lectures were conducted synchronously using multiple devices and applications. A laptop was connected to an external monitor to provide ample teaching space. The external monitor was restricted to *Microsoft Teams*, an online workshop offering chat, video conferencing, and software sharing; this allowed Makayla to monitor student engagement and interact through an external microphone and webcam. An *iPad Pro* acted as the second monitor; this was screen shared with the learners through *Microsoft Teams*.

The purpose of the tablet was to demonstrate sketching skills and visually answer questions and comments synchronously. The digital sketching approach used differed from Case Study 10.2 due to the previous lockdown teaching experiences and through nonacademically published mediums, discussed previously, that students often experienced difficulty viewing traditional sketching demonstrations due to the presence and positioning of the teachers' hand. *Miro*, an online collaborative whiteboard, delivered the sketching demonstration lectures. *Miro* board permissions were set to mean the visitors (learners) could view and add comments to any area of the board. *Miro* timers were also used to ensure the lecture remained on track, and *Miro* timer music provided ambience during individual learner activities. Before each lecture, a board template was created containing five core areas: (a) an introduction to the lecture and the teacher, (b) a reminder of core slides from that week's lecture, (c) a blank area for demonstration, (d) an exemplar area, and (e) upload area. Each sketching lecture began with a reminder of the lecture, then sketch-a-long to build sketching skills, e.g. actions, faces, figures, emotions, scenes, etc., followed by a series of individual activities for application purposes. During the lecture, students were encouraged to use *Miro* to ask questions by placing comments next to the

10.3 Remote Sketching Case Studies

Fig. 10.3 Example interactive *Miro* board for demonstration of sketching components and icons for low-fidelity paper prototyping lecture. *Miro* Whiteboard on *Apple iPad Pro* using *Apple Pencil*. Makayla Lewis and Miriam Sturdee, 2021

relevant sketch or material (yellow speech bubbles). This helped Makayla to keep track of questions and their relevance during synchronous demonstrations. As an example, Fig. 10.3 shows an interactive *Miro* board demonstrating sketching devices, interactions, and gestures for storytelling and storyboarding lectures with a guest speaker.

Towards the end of the lecture, the *Miro* board elements were locked in the last 15 minutes, and permissions were set so that only the teacher could unlock the board. The learners were then reminded that the sketching lecture was a safe and supportive environment to share their creations. The *Miro* board permissions were changed to allow learners to upload their sketches to a predefined board area using their smartphone cameras and the *Miro* app. This allowed the teacher to view and provide feedback to students synchronously; it also offered the students to give constructive feedback to each other. Students who chose not to share their creations were asked to upload them to their group *Microsoft Teams* private channel to obtain feedback from the teaching team and their group.

Case Study 10.3: Remote Peer-to-Peer Sketching

As previously noted, sketching is often overlooked in many applications and disciplines (Chap. 1); it is often referred to as a "soft" skill, and as such, the direction is usually not provided in universities and adult learning institutions. In 2016, the we observed that those wishing to learn and practice sketching in HCI had limited opportunities in a fun, engaging, confidence building, and friendly sketching environment. The ongoing journey to provide this opportunity began at the ACM-sponsored conference NordiCHI 2016 and has continued ever since across many conferences, summer schools, events, and meetups.

Our courses aim to be "hands-on" and foster a learning-by-doing approach, as with the previous case studies in this chapter. We guide people from hands-on sketching to practical research contexts, with opportunities for extended practice, feedback, and creative thinking. This book exemplifies and expands on these courses, which have been iterated upon since their inception. Those who participated were asked to be open-minded and open to sketching exploration—and, as a result, it is hoped that they will leave with the confidence to begin to employ sketching in their own HCI education, research, and practice.

Our courses are always open to all comers and therefore directed towards academics (teachers and researchers), industry leaders, practitioners, students, and early career researchers interested in learning and improving their sketching skills. There are no prerequisites for attendance. All with interest are always welcome. Depending on the venue, our courses can average between 120 and 240 minutes in length and have anything from 15 to 60 students, depending on the platform and the size of the event. We ran three online peer-to-peer courses during 2020–2022 and followed many of the activities in the chapters you have just read and (hopefully) engaged with (Lewis & Sturdee, 2020; 2021, 2022a, b)!

The COVID-19 pandemic meant that many HCI conferences in 2020 and 2021 were cancelled, postponed, or moved to online only. As we had already successfully transitioned from in-person to online-only sketching lectures and workshops (as outlined in the sections above), we were enthused to continue teaching Sketching in HCI to a broader community. Thus, during the pandemic, we conducted three courses with HCI students, practitioners, and researchers worldwide. The courses occurred synchronously online across varying time zones, and we presented from the early hours of the morning, late afternoon, and late evening to night to provide maximum community reach.

A digital sketching setup was used by both of us to better support multiple views and interactivity and to allow international audiences to follow and sketch-a-long directly via a *Miro* board if the video conferencing platforms (*Zoom*, *WebEx* (W8), and *Microsoft Teams*) were unclear, either due to learners' technical issues or low Internet connectivity (bandwidth). *Miro* was used to deliver the courses and support students' presence and engagement, enabling content locks, timers, comments, emojis, and timer music (Figs. 10.4 and 10.5).

The online synchronous courses, although different in delivery from the pre-COVID courses, had the same learning outcomes, demonstrations, sketch-a-longs, hands-on activities, and two facilitators. Together we developed an extensive *PowerPoint/Google Slides* deck, 126 slides (at the time), of which 20% were made visible to students, and the remainder were teachers' timings and prompts (Fig. 10.6). One of us was responsible for screen sharing the slides, but we took equal turns to present theory and examples and provide video conferencing chat moderation. We also pivoted screen-sharing sketching demonstrations on the *Miro* board and

10.3 Remote Sketching Case Studies

Fig. 10.4 Online learning sketching setup in a quiet, well-lit, location: 27" monitor, external microphone, external webcam, external keyboard, external mouse, and *Wacom Cintiq One* with pen on a laptop stand. *Miro* Online Whiteboard on *Apple iPad Pro* using *Apple Pencil*. Makayla Lewis and Miriam Sturdee, 2021

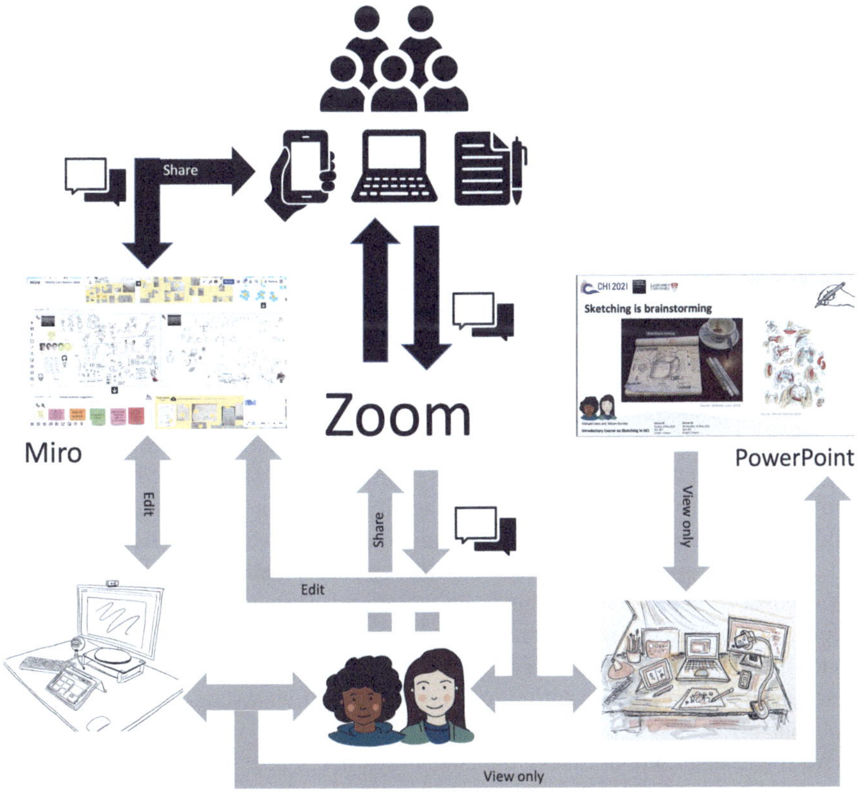

Fig. 10.5 ACM CHI 2021 delivery setup. *Microsoft PowerPoint* and *The Noun Project* icons, Makayla Lewis and Miriam Sturdee, 2023

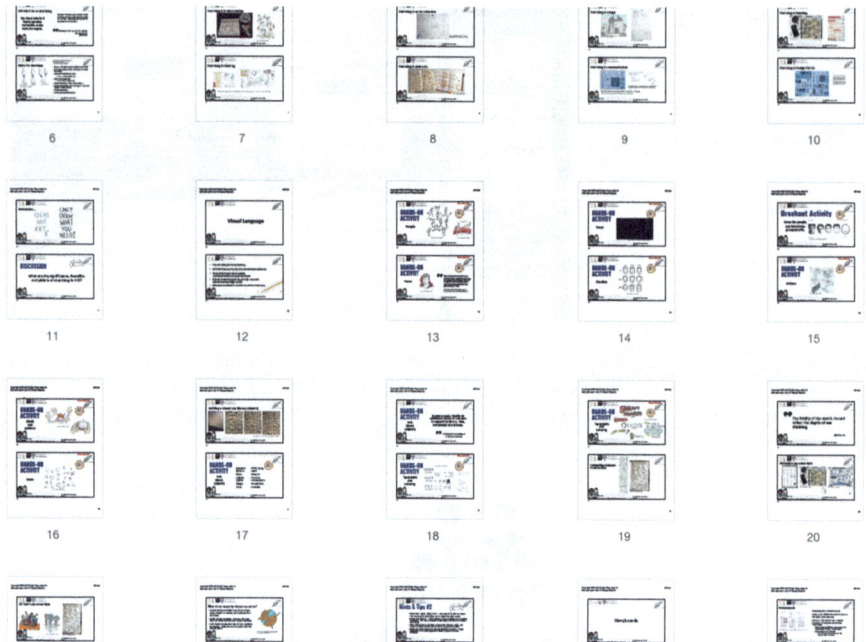

Fig. 10.6 Extract of CHI 2023 "The Joy of Sketch" course slides, *Microsoft PowerPoint*, Makayla Lewis and Miriam Sturdee, 2021

moderation of *Miro* comments and emojis for feedback. To further support the dual delivery, the course's *Miro* board underwent alterations from course to course to support practical engagement better.

In the peer-to-peer context, the most significant differences between online and in person were the fear of digitally sharing images—especially about peer judgement—in a limited community building space. This meant they were less trusting and thus reluctant to share their sketches. Participants were engaged less consistently, perhaps due to demands of working from home or distracting spaces, e.g. children and pets were often seen and heard. Sometimes, everyday household noises and external factors, such as deliveries, traffic, and aeroplanes, were present. Despite this, feedback via social media during and after post-courses was positive: "Despite it being 4:30 in the morning here, having lots of fun at the #chi2021 course 'Let's Sketch! A Hands-on Introductory Course on Sketching in HCI'" and "Such fun sketching at virtual chi! I want more hands-on virtual workshops :) thanks for the cool course!".

We found that humour during the course was necessary; we tried to be fun and engaging but found "the room" is easier to read when in person; people often had their webcams and microphones off—thus, receiving visual and auditory feedback was difficult for us. We also found it much harder to cover the learning objectives, discovering that a three-quarter-day sketching course would be far more physically and emotionally draining when hosted online rather than in person. However, the

10.3 Remote Sketching Case Studies

online delivery setup did allow us to demonstrate and collaborate with sketching simultaneously in the same *Miro* area, an aspect not possible when sketching using analogue tools (flipcharts and whiteboards have physical space constraints). Furthermore, there is now a permanent online record and textual feedback online, meaning learners can revisit their work and the course in a way that was impossible with our in-person offerings (although some students were uncomfortable with this aspect and so deleted their shared sketches post-course). Finally, we identified that online courses are cheaper to run and make it easier to meet and teach with people worldwide who might not have attended our conference courses previously due to their in-person format. So, our community reach during the pandemic was the highest ever observed over the 6 years since the inception of our Sketching in HCI work. Although the online delivery of sketching education was well-received, we determined that in-person courses can be beneficial, as it is easier to circulate the space and offer feedback and encouragement in the minute.

Case Study 10.4: Hybrid Sketching at CHI
The COVID-19 pandemic significantly impacted the way in which people communicated, learnt, and worked. This led to a rise in hybrid models that offered greater flexibility by combining elements of remote and in person at designated locations. For the us, the hybrid models for which they are now working influenced their decision to leverage technology to bring sketching education to a wider audience. It was hoped that extending the course beyond the physical limitations of a conference venue would offer an opportunity to deliver a versatile and inclusive format that can accommodate a broader range of participants (to participate from anywhere thus eliminating the need for travel expenses and logistical challenges).

The "Joy of Sketch" course occurred at a physical venue in Hamburg, Germany, at the ACM-sponsored conference CHI 2023 (Lewis & Sturdee, 2023). In Room X07, a well-lit room with large tables and a projector, in-person participants gathered with their pens and paper. The course was synchronous in format, thus also live-streamed using the *Zoom* conferencing platform. Online participants were asked to turn on their webcams and microphones, lay out their sketching materials, and have their smartphones on hand (e.g. Fig. 10.4). All participants were asked to download the *Miro* app onto their smartphones; online participants were also asked to open *Miro* in their chosen browser. Makayla directed in-person and online participants to the course *Miro* board to upload their creations, and a "how-to" demonstration was provided (Fig. 10.7).

Makayla was tasked with screen sharing the slides, time management, music, and online engagement in activities. Miriam focused on in-person engagement and conference organisers and IT support. We played an equal role in delivering theory and demonstrations. *Microsoft PowerPoint* was used to present the slides and demonstrations that occurred live on *Miro* (Miriam using *Microsoft Surface Go* and *Pen*; Makayla using *Apple iPad Pro* and *Pencil* (Fig. 10.1). "Show and tell" occurred in two formats; those in person were asked to share their work with those at their table and upload to designated areas of *Miro*, whereas the online attendees were asked to only upload to *Miro*. This division in delivery that centred around a digital

Fig. 10.7 Hybrid *Miro* board from ACM CHI 2023 conference course "Joy of Sketch". *Miro* Online Whiteboard on *Apple iPad Pro* using *Apple Pencil* and photographs of analogue sketching tools (participant sketches are blurred for anonymity). Makayla Lewis and Miriam Sturdee, 2023

whiteboard ensured engagement, collaboration, knowledge sharing, and learning was as equal as possible (Fig. 10.8). The course gained significant popularity because hybrid engagement has become more prevalent thus limited issues occurred (Lewis & Sturdee, 2023). One online participant commented in the live chat, "This is the only hybrid course at CHI 2023 that I could find; it's nice to be included".

A key learning point for us was that it is crucial when co-facilitators are not physically present to ensure effective planning, communication, and hybrid treatment and engagement amongst attendees. Miriam often asked how those online were doing and whether they had questions or wanted to share and/or discuss their creations, whilst Makayla regularly reminded attendees to upload their sketches to *Miro* and comment on each other's creations. This adaptation in teaching and

10.4 Hands-On Activities

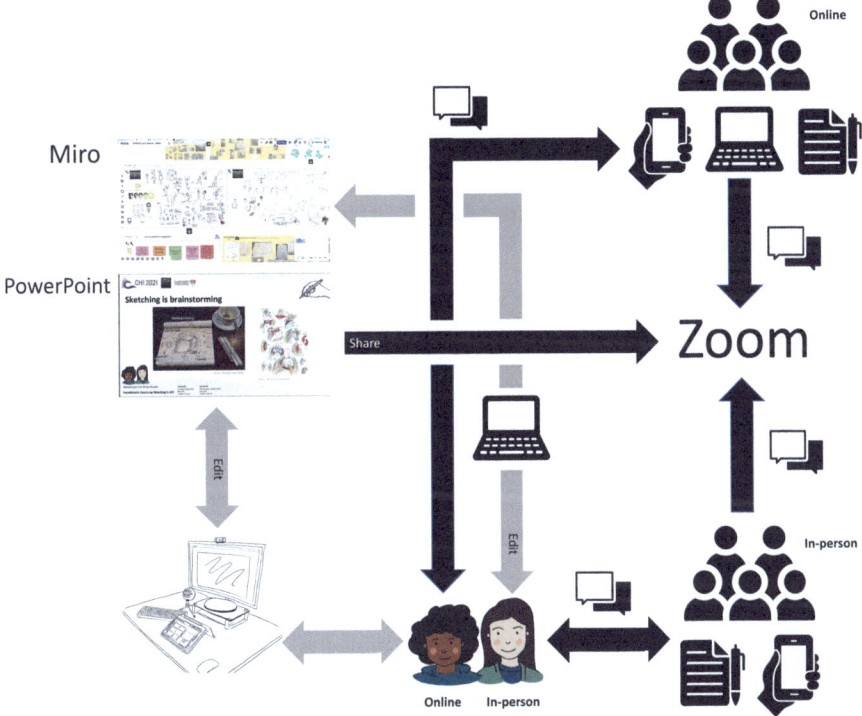

Fig. 10.8 ACM CHI 2023 hybrid delivery setup. *Microsoft PowerPoint* and *The Noun Project* icons, Makayla Lewis and Miriam Sturdee, 2023

learning was scary at first; "humour" became a key support when technology did not play ball. Regular breaks where in-person attendees visited a refreshment stand outside of the room and online attendees were asked to take a screen break and stretch meant that we could regroup and adjust plans to ensure fair treatment for all participants. Going forward we have agreed that hybrid, although stressful and requiring a vast amount of planning, was the future of sketching education at conferences.

10.4 Hands-On Activities

Activity 10.1: Sketch Your Online Setup (Individual Activity)
Learning objective—Sketching digital and analogue devices within context of physical spaces
Time—20 minutes
Materials—Pen, paper, and smartphone, *Miro* (hybrid); digital sketching device, *Miro* (online only)

Procedure:

- Each person is asked to draw the research or topic of interest on paper using a back pen. Facilitator informs participants to not write words in their sketch; it should be clear and concise as possible (10 minutes).
- Each person completes their setup sketch and uploads it to the board (5 minutes).
- Each person should then introduce themselves and describe their current and preferred setup and what works/does not work when sketching remotely.

This is a useful activity for more experienced sketchers, or those who are confident diving in at the deep end! For novice sketchers, pivot some of the warm-ups to a digital format. In fact, many of the activities in this book can be moved to online formats with a little preparation.

Activity 10.2: Without Words—Digital Edition (Group Activity)
Learning objective—Support hybrid introductions and communication
Time—20 minutes
Materials—Pen, paper, and smartphone, *Miro* (hybrid); digital sketching device, *Miro* (online only)
Procedure:

- 5 minutes—Each participant is asked to draw the research or topic of interest on paper using a black pen or digital device. Facilitator informs participants to not write words in their sketch; it should be clear and concise as possible.
- 5 minutes—Each person completes their Without Words sketch.
- 5 minutes—Facilitator asks the participants to take a photograph of their sketch and upload it to a *Miro* designated area.
- 10 minutes—Participants are instructed to use *Miro* Post-it notes to anonymously guess each other's research or topic of interest.
- 5 minutes—Using different colour *Miro* stickers, the authors of the sketch are to respond to each response with a thumbs up or thumbs down.

Following completion of this task, participants can discuss what representations were more successful, or liked the most. They could also discuss how the communication could be improved and what new icons or concepts could be shown. Figure 10.9 shows one of our Without Words *Miro* tasks, alongside our own contributions (left).

Activity 10.3: Virtual Visitation (Group Activity)
Learning objective—Navigate digital environments and sketch "on location"
Time—30 minutes
Materials—Pen, paper, and smartphone, *Miro* (hybrid); digital sketching device, *Miro* (online only)

10.4 Hands-On Activities

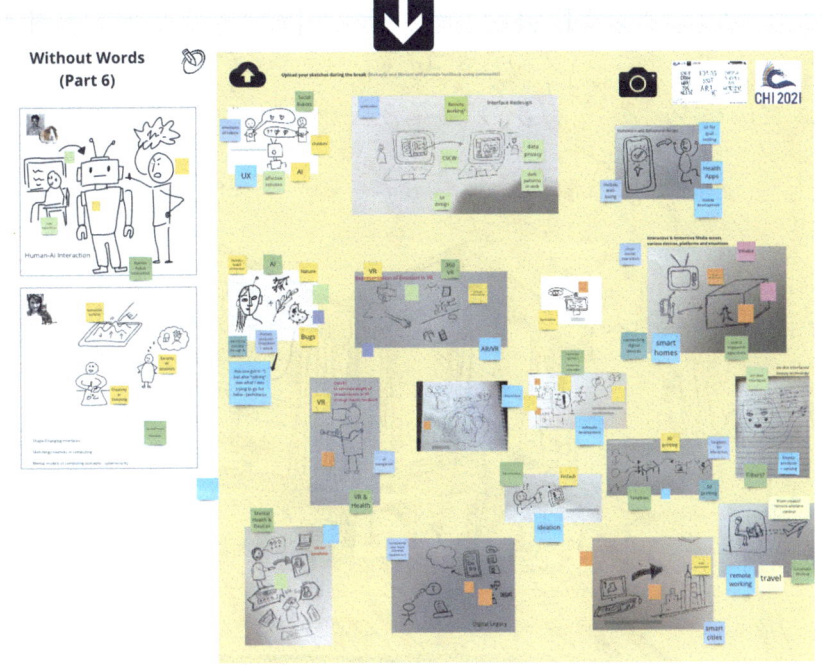

Fig. 10.9 "Without Words" task from the ACM CHI Sketching HCI course. *Miro* Online Whiteboard on *Apple iPad Pro* using *Apple Pencil* and photographs of analogue sketching tools (participant sketches are blurred for anonymity) 2021

Procedure:

- The facilitator or teacher should choose a virtual gallery or museum website (e.g. the Smithsonian) or visit one of the many places available on *Google Arts and Culture* (W9).
- All participants should be directed to the same website and invited to explore the interactive experience.
- The first task is for participants to pick *one* location and sketch the scene, the room, building, or general environment (5 minutes) (see Fig. 10.10).
- All participants should upload their image to *Miro* and label it with their name and the location.
- Next, participants should all choose a venue detail (or several details)—or if visiting a gallery or museum, then choose artworks or artefacts, and sketch those (see Fig. 10.11 for an example) (10 minutes).
- Finally, participants should create an overview, or vignette (Fig. 10.12), of the overall experience from the visit (15 minutes).
- All participants should then upload their images to the *Miro* and chat about the experience and their favourite exhibits or areas in the virtual place or gallery/museum.

Fig. 10.10 Virtual visit to the Sankeien Garden in Yokohama on *Google Arts and Culture*, sketch of one of the buildings as part of the ACM CHI Sketching Special Interest Group (Sturdee et al., 2021), Fineliner pen and marker on paper. Miriam Sturdee, 2021

Fig. 10.11 Virtual sketchnote during COVID-19 UK lockdown "Hirshhorn Museum Sculpture Garden". Fineliner pen and marker on paper. Makayla Lewis, 2020

Fig. 10.12 Overall interpretation of the Sankeien Garden in Yokohama on *Google Arts and Culture* as part of the ACM CHI Sketching Special Interest Group (Sturdee et al., 2021). Fineliner pen and marker on paper. Makayla Lewis, 2021

References

Books and Papers

Lewis, M., & Sturdee, M. (2020). *A hands-on introductory course on sketching in HCI: research practice and publication in HCI*. Kingston University.

Lewis, M., & Sturdee, M. (2021, May). Let's sketch! a hands-on introductory course on sketching in HCI. In *Extended abstracts of the 2021 CHI conference on human factors in computing systems* (pp. 1–4).

Lewis, M., & Sturdee, M. (2022a). Curricula design & pedagogy for sketching within HCI & UX education. *Frontiers in Computer Science, 4*, 826445.

Lewis, M., & Sturdee, M. (2022b, April). Take a line for a walk! A hands-on introductory course on sketching in HCI. In *CHI conference on human factors in computing systems extended abstracts* (pp. 1–4).

Lewis, M., & Sturdee, M. (2023, April). The joy of sketch: A hands-on introductory course on sketching in HCI and UX within research, practice, and education. In *Extended abstracts of the 2023 CHI conference on human factors in computing systems* (pp. 1–4).

Sturdee, M., Lewis, M., Spiel, K., Priego, E., Fernández Camporro, M., & Hoang, T. (2021, May). SketCHI 4.0: Hands-on special interest group on remote sketching in HCI. In *Extended abstracts of the 2021 CHI conference on human factors in computing systems* (pp. 1–4).

Websites

W1 Microsoft Teams – www.microsoft.com/en-gb/microsoft-teams/group-chat-software
W2 Google Drive www.drive.google.com/drive
W3 Zoom – www.zoom.us/
W4 Google Meet – www.meet.google.com/
W5 Miro online digital whiteboard – www.miro.com/
W6 Mural online digital whiteboard – www.start.mural.co/
W7 FigJam – www.figma.com/figjam/
W8 WebEx – https://www.webex.com/
W9 Google Arts and Culture, Sankeien Garden, Yokohama – www.artsandculture.google.com/partner/sankei-en

Further Reading

Lewis, M., Toselli, M., Baker, R., Rédei, J., & Ohlenschlager, C. E. (2022, June). Portraying what is in front of you: Virtual tours and online whiteboards to facilitate art practice during the COVID-19 pandemic. In *Proceedings of the 14th conference on creativity and cognition* (pp. 350–363).

Chapter 11
Applying Sketching in Research and Practice

11.1 Destination, Sketching!

You've made it so far! We hope that you are feeling more confident in your sketching practice and are ready to take things to the next level. By next level, we mean bring your sketching out of the learning zone and into your daily life, whether in work or play (because, let's face it, sketching is also fun!).

We are both researchers, who not only specialise in teaching people how to sketch across HCI, UX, and computer science contexts, but we also use sketching in our own research. By exploring some of the work we have done and places we have used our skills, hopefully you can see the breadth of application for sketching both in research and academic arenas and also in industry roles. Some of you may have your eye on a UX job, and sketching is a valuable skill in that domain, so don't be afraid to advertise the fact that you are a confident sketcher.

We primarily cover individual research and practice within this chapter, as working with other people in the sketching domain is a vast topic in itself and therefore deserves its own chapter—but many of the sections in this chapter can also be applied to the next (e.g. networking and sketch analysis). We therefore suggest you work through these chapters together, picking sections and activities that are most relevant to your own interests.

This chapter outlines some of the practical ways we have both used sketching in our work (and sometimes outside of work in networking contexts) and hopefully will inspire you to utilise the skills you have gained from the pages of this book to embark on your own professional research and practice with sketching.

By the end of this chapter, you should be able to:

1. Identify opportunities to sketch in networking contexts.
2. Create and manage your digital sketching footprint.
3. Know how to utilise your sketching skills in group contexts.

4. Understand how to use sketching in research methods and outputs.
5. Continue improving your practice using observational sketching.

11.2 Sketching and Networking

Sketching is a wonderful way to break down boundaries between people, and you can think of it as everything from a conversation starter or dialogue to a gift. The rise of social media has made it easy to share images, and even platforms that are primarily text based have the capacity to add imagery—imagery also can increase engagement with a post!

Sketchnoting someone's research or industry talk (whether in person or online) is a great way of engaging with that person and their research group or company; it shows you were actively listening! But of course if nobody sees your sketchnote… Did it even happen? You don't have to post every sketch you make, especially if you aren't pleased with the final result, but by sharing publicly and tagging the relevant people, you can gain a presence online. In our experience, we get the most engagement with posts on platforms such as *X* and *Instagram* when we share imagery—but don't forget the AltText!

Sketching socially can also be useful for networking. By socially, we mean sketching for yourself and your own progress, and sharing the results with your friends and peers. As an example, we both took part in a fortnightly "Drink and Draw" activity for a year during lockdown, run by the London-based Gosh Comics (W1) and *Broken Frontier* (W2) (hashtag #GOSHBFDD if you wish to see the many wonderful drawings by the comic artists and sketchers who regularly take part!).

This online events consisted of three rounds of quick-fire sketching, around themes or prompts chosen by guest artists. Each round gave people 30 minutes to respond to the given prompt, after which there was 15 minutes to upload and share on *X*. The zany prompts and quick-fire nature of the event are the perfect way to sketch loosely and without fixating on minor details. Figures 11.1 and 11.2 are examples of our work and the theme which inspired them—beware—some of these are peculiar!

So why are we mentioning this here?

Well, even the sketches we produced for this fun and social activity gained traction online, helped us gain followers, and even start new research collaborations! The very act of sketching is a fun and social thing; people enjoy looking at images that aren't work related, and this can lead them to see what else you do. Online and social media may be a great way to share your work (see next section for more on your digital sketching footprint), but sketching for networking is not only an online activity. Simply sitting and sketching at an event in person can invite conversation from others and help you get to know people at events where you might be the only person from your university or business. Sketching is also a great way of engaging

Fig. 11.1 #GOSHBFDD Drink and Draw: (right) rituals, (centre) puke banter, and (left) hell phone. *Procreate* App on *Apple iPad Pro* using *Apple Pencil*. Makayla Lewis, 2023

Fig. 11.2 #GOSHBFDD Drink and Draw: (rom left to right) prompt 1, Journey into Mystery; prompt 2, House of Secrets; prompt 3, Weird Science. Fineliner pen and marker on paper. Miriam Sturdee, 2022

with workshops, as there are often hands-on tasks which use pen and paper such as brainstorming on flipcharts or whiteboards. Being a person who is confident enough to help others visualise their ideas via sketching not only helps with group cohesion but also means the workshop outputs are richer.

Being known as "the person who sketches" in your department or school can lead to fruitful collaborations, and you may find your skills are often in demand—people who have this skill are valuable. However, beware; don't take on too much extra work without attribution or value to yourself; never sell yourself short! Sketching is still work and should be appreciated for the time and effort put into it.

Practical Application Tips

- If you don't have the time to sketchnote at an event or conference, try making little portrait sketches of the people around you. If you are pleased with the results and know their name, tag them on social media!
- Add small quotes to vignettes or portraits to show the context they were created in.
- Remember to sign and date your sketches before you photograph or scan them; then when they are shared further, you can still be attributed.
- Take photographs of your sketches in the best light possible, and make use of filters to clear noise and make the images "pop".

Try this technique for yourself by completing Hands-On Activity 11.1.

11.3 Digital Sketching Footprint

Creating a space online (sometimes referred to as a digital sketching footprint) that is your own, is essential to disseminating yourself and your work. A little creative area amongst the millions of websites and social media accounts can help to:

- **Take control**—It's your digital footprint; no one can decide what you add to your space; you can regulate your content, design, and purpose to meet your needs and preferences.
- **Network**—An online presence can help you to connect with others in and outside your area of interest; it can help you build new connections. This is how Miriam and Makayla met online in 2014, met in person 2016, and now are writing this book all these years later.
- **Build a community**—An audience where you can discuss your work (and their work) e.g. the #SketCHI hashtag on now *X*, which we created and often upload, or #TodaysDoodle hashtag on *X* and *Instagram* that Makayla joined 9 years ago and still uploads to (Fig. 11.4).
- **Visual diary**—Creating a visual record of your practice that you can reflect on (i.e. return to see things that worked and did not), can be a positive reminder of the growth of your sketching skills. If you look on *Flickr* (W3, W4), you will find Makayla's first sketchnote; it is questionable (hard to read, poor structure, and inconsistent), and she will not give you the link. Happy hunting!
- **Portfolio of your work**—Share your skills, visual style, and previous sketches as this can lead to collaborations; it can also boost your credibility.
- **Long-term archiving**—Your digital footprint can serve as a long-term archive for your sketching; thus, if you misplace an illustration, you know where to find it quickly and redownload or share the link. Most of the images in this book by Makayla were downloaded from her website and *Flickr* (Fig. 11.4), she has no idea where the originals reside! Thus, ensure you upload high-quality photos to a platform that allows you to download your images in the future.

Makayla has had a personal website and social media presence since 2008 when she began her PhD (see Fig. 11.3); it has led to a growth in audience, connections

11.3 Digital Sketching Footprint

BY DAY...

Dr. Makayla Lewis (She/Her) has a Ph.D. in human-computer interaction from City University London. Her key interests are human-computer interaction, user experience design, design thinking, human factors in cybersecurity, artificial intelligence, smart money, and sketching in HCI.

Currently, Makayla is a Senior Lecturer in Computer Science (User Experience Design) at Kingston University London; and a Design Associate at Design Council in the UK. Makayla achieved Fellowship of the Advance HE (FHEA) in 2022 for teaching and learning in higher education

Fig. 11.3 Makayla Lewis' website homepage: www.makaylalewis.co.uk

(friendships and professional—and a mixture of both), and credibility. To ensure the success of your digital footprint, we recommend the following:

- **Personal brand**—Your visual presence should be consistent, e.g. avatar, username, theme, message, and language style, and the goal should be clear and consistent across websites, platforms, and social media.
- **Personal domain**—Owning a personal part of the web can be a memorable way to represent yourself and for people (potential collaborators) to find you. For this to happen you should ensure your domain name is logical, e.g. www.makayla-lewis.co.uk (W5). As your aim is to increase your visibility in search engine results, thus supporting discovery.

Fig. 11.4 Extract from Makayla Lewis' *Flickr* Gallery for TodaysDoodle uploads. This is an ongoing project and is regularly updated

- **Image copyright**—Identify the level of the copyright you want to use for your sketches, and ensure it is present and consistent on all your digital graphics. This will ensure your sketches are respected in terms of ownership, manipulation, and dissemination. Makayla believes copyright is essential to maintaining your brand.
- **Content**—Do not only share the sketches you have created; you want your viewers to engage with your content. To do this, they need prompts or something to respond to, e.g. work in progress, a detailed description of the sketch, thoughts regarding the process, expertise or tips regarding the process, and/or reflections, i.e. what could be improved.
- **Content insights**—Upload your sketches to social media platforms (apps and websites) that offer analytics, e.g. *WordPress*, *Instagram*, *Flickr*, etc. This will allow you to track visitor behaviours and determine what your audience is interested in, thus providing direction for improvement and/or future content.
- **Contact information**—Your website and social media should allow others to contact you; this can be a contact form on your website, open DMs, or links to your contact webpage on social media.

Warning: do not share your email, telephone number, address, or personal information publicly. Your safety is a priority, as anyone online can access public data. People online are strangers; we all remember "stranger danger" when we were younger (Figs. 11.4 and 11.5).

Whether for personal or professional purposes, a digital footprint can be a powerful tool for self-expression, communication, and achieving your goals as an expert sketcher in HCI whether a student, educator, practitioner, or researcher.

11.4 Sketching as a Research Method

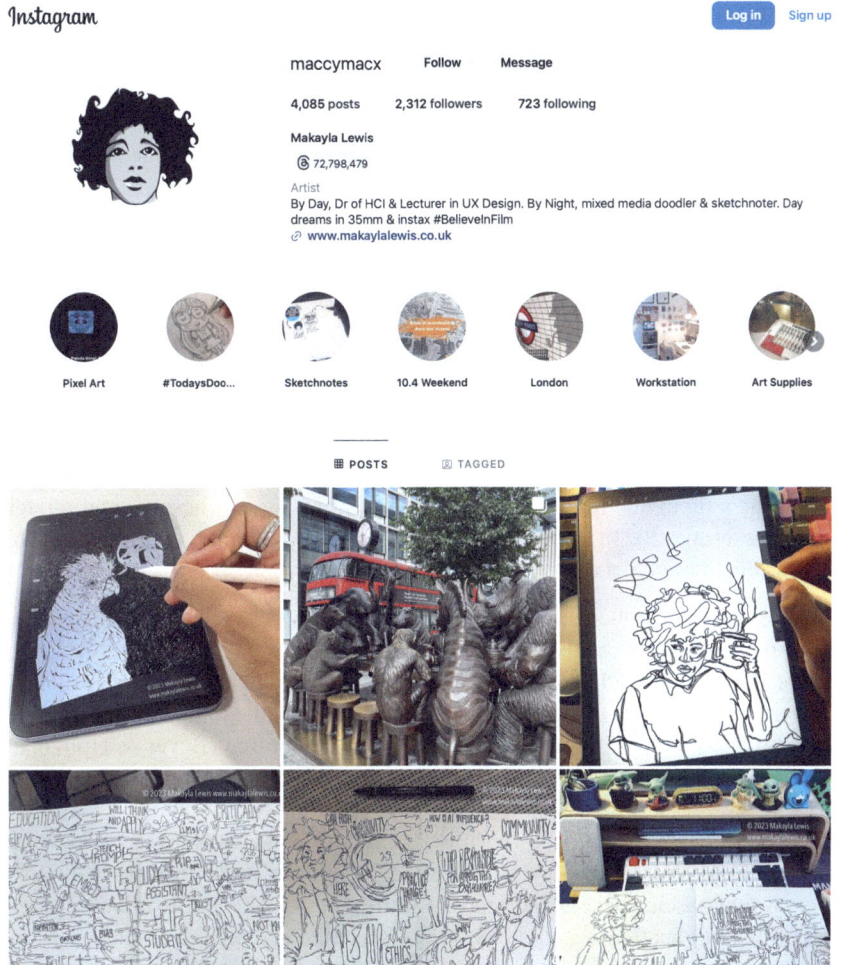

Fig. 11.5 Extract from Makayla Lewis' *Instagram* with collects together doodles, sketches, sketchnotes and illustrations. Alongside creative events and gallery/exhibitions/installations across London and beyond

11.4 Sketching as a Research Method

Sketching has long been celebrated as a research method in art and design but has long been relegated to the sidelines within other disciplines, despite its value and wide uptake! Books such as Sketching User Experiences (Buxton, 2010) and its workbook (Greenberg et al., 2011) helped foreground the habit, but this was within very specific contexts, and sketching in HCI is so much more than just user experiences. We cover many of our thoughts about the theory behind this in the first, introductory chapter, but here in this section, we suggest a few practical examples for research.

- **Sketching as a primary response**—This is where *you* (the researcher) are using sketching to create something new from a situation, for example, doing visual ethnography, or using sketching as a form of "reportage"—documentary of events in your own subjective style.
- **Sketching as a secondary response**—This applies to sketches created by the researcher to represent textual data (perhaps an audio transcription or reaction to a video). These could be visual responses, or explanatory, diagrammatical, or illustrative.
- **Sketching as interrogation**—Here, we imagine sketching as a way for the researcher to work through and with data, as part of thinking through the research process. Data could be both textual or visual, or even audible or tangible! This can also be done collectively.
- **Sketching as data collection**—The sketches of others can tell you a lot about someone's thoughts, feelings, and perceptions of a topic. These mental models (Sturdee et al., 2021a) can be analysed as data from various participants. We cover this in more detail in Chap. 12.

Where we have collected sketch-based data, either personally, from existing online sources, or from active participants in research studies, we can then think about analysis. Both authors have been exploring visual research methods in HCI for several years and have adapted existing qualitative methodologies (e.g. thematic analysis (Clarke et al., 2015; W7) and affinity diagramming (Beyer & Holtzblatt, 1999; Harboe & Huang, 2015) to help interrogate the sketch data and form theories and ideas.

In Figs. 11.6, 11.7, and 11.8, we show the process of sketch analysis for sketches created to represent dreams about technology, from our research work with John Miers on subconscious user experiences (Sturdee et al., 2022). The initial research stage of this publication was the creation of sketched and artistic responses to the three researchers' dreams about technology. These sketches formed the primary output of the pictorial paper and were presented as full-page images.

Following the final sketches being complete, we then uploaded them to a *Miro* board and visually "coded" the images, based on our responses and subjective analysis of the images. For example, the top right image in Fig. 11.6 has one response "crushed in the crowd" and bottom left has "imminent danger". Responses using this process must be instinctive, and not overly considered, so as to capture the initial response. This part of the process is similar to the familiarisation and coding process in thematic analysis (see Emeline Brule's article (2020) for a good overview of thematic analysis specifically for HCI).

Once the reactive stage has been completed, the sticky notes (or physical Post-its—you can also do this using analogue sketches) are read and reread (familiarisation, part 2 (Brule, 2020)) by all researchers and grouped during a discussion phase. This discussion is an important part of the analysis. The first grouping may not be

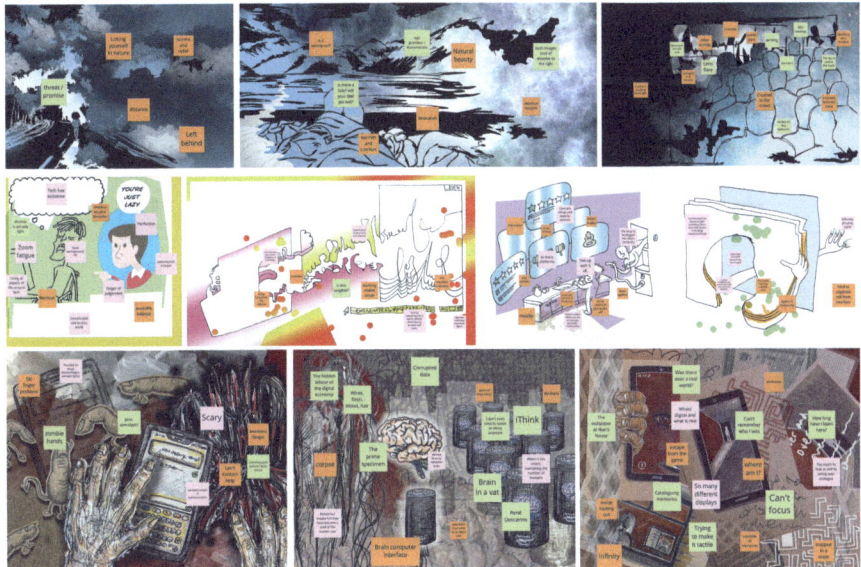

Fig. 11.6 Visual analysis using *Miro* and sticky notes on sketches from Sturdee et al. (2022): top, sketches by Makayla Lewis (*Procreate* App on *Apple iPad Pro* using *Apple Pencil*); middle, sketches by John Miers (*Procreate* App on *Apple iPad Pro* using *Apple Pencil*); bottom, sketches by Miriam Sturdee (mixed media on paper)

the last however! Take a break, and return to the analysis with fresh eyes. When you do, consider the following questions:

- Were there any notes that didn't seem to fit anywhere?
- Do the overarching themes make sense in the context of the research?
- What do they tell us about the topic?

The final version (Fig. 11.8) is organised and grouped into 3 main themes, and 20 sub-themes. Connectors are used to link up concepts that are related, to show that the sub-themes are not exhaustive and some items relate to other spaces. Despite iterative sorting, we remained with a "Misc" section; although revisiting this now, perhaps we can see other connections as we have developed our practice!

The final themes, sub-themes, and examples can finally be described in detail (if presenting your work textually) or visualised in detail if you are working on a primarily image-based output (see Sect. 11.4). For another example of visual analysis of sketches see Sturdee et al. (2021a) where participant sketches are analysed in a similar way and tell us about mental models of cybersecurity. Both authors also created a doodle-based research project where we sketched during online meetings and analysed the resultant imagery (see Figs. 11.9 and 11.10) (Lewis, 2023). Furthermore, we then used the analysis to create textual descriptions and sketched responses for the body of the publication (see Fig. 11.11); this shows how these visual methods continue to develop this rich way of working with sketches and give evidence as to its efficacy.

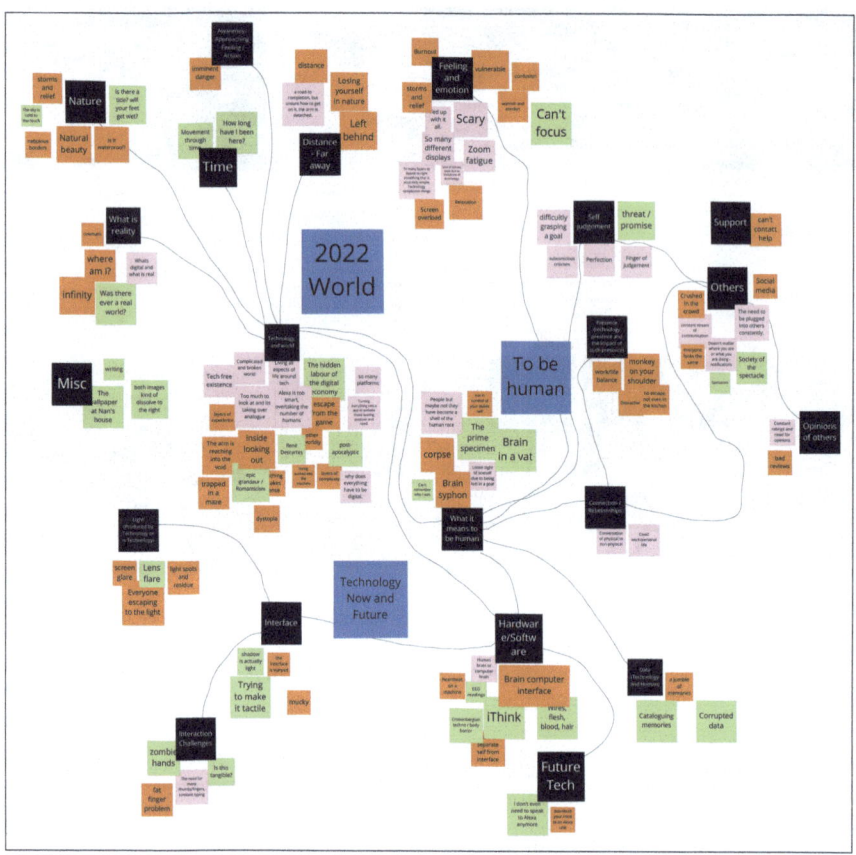

Fig. 11.7 Initial grouping by affinity for raw sticky note data. *Miro* Desktop App, 2022

The methods described here was created by the authors to fill a gap in qualitative analysis when working with sketches and may not be exhaustive. There are many ways to analyse and write about (or visualise) sketched data, not all of them formalised. There may also be new ways of creating and working with sketch-based data, so we invite you to explore this area!

Practical Application Tips

- Sketching can be used at all levels of the research process.
- Any sketched data can be examined and analysed, whether created by the researchers or participants.
- Visual data can offer a richer interpretation of a theme than text and allows for ambiguity.
- One size does not fit all; adapt existing analysis and process to fit your research questions and outputs.

11.5 Publishing Your Sketches

Fig. 11.8 Finished grouping and presentation of themes and sub-themes. *Miro* Desktop App, 2022

- Try to do initial coding or analysis of sketched data with other people, but if the analysis is purely subjective, that's ok—you can state this in your output.
- Your final response can also be sketched—a full textual description is not always essential in visual publication.

Try this technique for yourself by completing Hands-On Activity 11.2.

11.5 Publishing Your Sketches

Sketches are a vital part of any design process, yet HCI has traditionally been shy of celebrating processes and imagery in archival academic outputs, partly because of a demand to keep page lengths down and partly because the written word is still the primary choice for academic output. Luckily, this is changing. The page length issue (which should have been phased out a long time ago when digital reproduction

Fig. 11.9 Sketched and textual response to the initial image analysis from the "Doodle Away" paper (Lewis et al., 2023). Fineliner pen and marker on paper. Miriam Sturdee, 2022

became the primary output in academic journals and conferences (Sturdee et al., 2018)) has been solved by instead asking for word count limits in many places.

Further, many academic venues now are accepting a new kind of archival publication called *pictorials* (Blevis et al., 2015). A pictorial is a fully archival (official record of peer-reviewed research) publication where the visual narrative is centred and celebrated. It allows authors to provide examples of visual process and carefully curated imagery to support the narrative of a research project or argument. Pictorials

Fig. 11.10 Sketched and textual response to the initial image analysis from the "Doodle Away" paper (Lewis et al., 2023). *Procreate* App on *Apple iPad Pro* using *Apple Pencil*. Makayla Lewis, 2022

were first developed and established by a team at Simon Fraser University and accepted at ACM DIS 2014 (ACM Designing Interactive Systems), although a rare few examples of primarily pictorial work existed within the "normal" conference outputs prior to this. A number of ACM (Association for Computing Machinery) HCI venues now accept pictorials, and some design-focused venues. Even HRI conference (Human-Robot Interaction) is embracing the pictorial revolution, although that is still confined to the "alternative" extended abstracts track. Both of us have

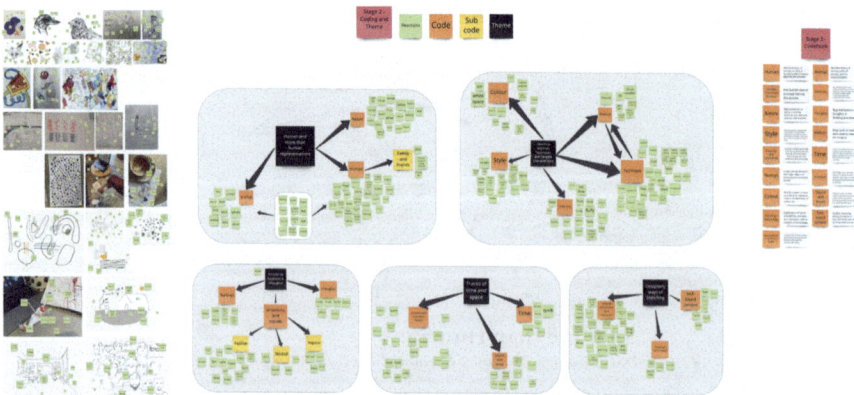

Fig. 11.11 Analysis of initial sketches from the "Doodle Away" paper (Lewis et al., 2023). *Miro* Desktop App, 2022

crafted and had multiple pictorials accepted for publication, together and separately (e.g. Lewis et al., 2022a, b, 2023; Sturdee et al., 2021b, 2022).

"Alternative" tracks at conferences are an ideal place to look for novel ways of using your sketches as a research output—with some even accepting physical artefacts and "critiques" (W6), but why not try a full pictorial publication if you are doing visual-focused research? There is also the opportunity to try and make visual narrative a bigger part of traditional textual conference and journal outputs, but often reviewers in these areas struggle to ascribe the appropriate value to visuals, and sketches in particular can be seen as "rough" or unfinished; however they are a valuable accompaniment to any paper describing process or as examples of output! We have engaged a lot with alternative tracks (alt-conference proceedings, workshops) and styles over the last few years (e.g. Lewis et al., 2022a, b, 2023; example, Figs. 11.12 and 11.13).

Practical Application Tips

- Ask yourself; are you conducting research with a predominantly visual theme? Or is it better described in a traditional textual format?
- Become familiar with layout tools and templates for the pictorial formats; if you cannot get access to *Adobe InDesign*, *PowerPoint* allows many of the same features.
- Don't focus on traditional section headings and introductory/discussion-based sections—could your imagery and annotations tell the story instead?
- Explore different layouts and colour schemes; pay particular attention to the flow of the visual narrative in the research story context.
- Annotate and elaborate on your images as needed.
- Remember to add AltText or AltNarrative to your final PDF!

Fig. 11.12 Extract from Makayla's pictorial from workshop at ACM SIGCHI conference 2022

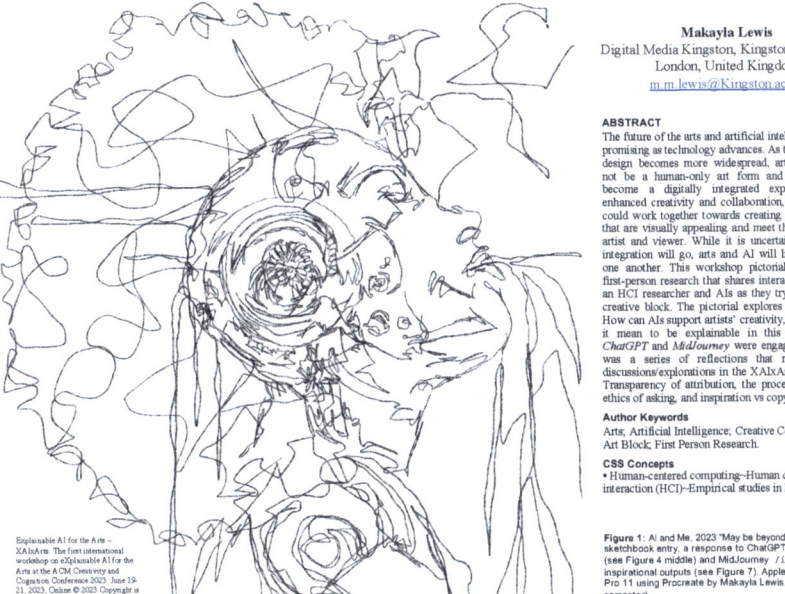

Fig. 11.13 Extract from Makayla's pictorial from workshop at ACM Creativity and Cognition conference 2023 (Lewis, 2022)

11.6 Visual Abstracts

Visual abstracts are becoming a larger part of academic publishing, yet rarely do researchers have the skills to produce meaningful imagery to summarise the scope and range of their papers. Instead, you will likely see the same graphs as are used in the analysis, or a simple *PowerPoint* flowchart, perhaps some icons. There are a number of publications examining the precise phenomena, and even discussing how sharing them on social media leads to higher engagement with research (Ibrahim, 2018). The main HCI conferences have not adopted the visual abstract yet, but a number of larger publishers strongly recommend their inclusion for journal articles.

On the flip side, a visual abstract does not have to be purely a research output; it can also be a guide for you in your own research and writing, helping you summarise important outcomes from a paper and creating a memorable device for revision in case of exams or presentations. Obviously, we advocate for reading all of a paper, not just the abstract, but as a starting point, the paper's textual abstract can tell you the main points and offers an easy introduction to making the research visual (Fig. 11.14).

The first thing to do is get a marker or highlighter and pick out the important points and things that will encourage visual imagery. Remember all of those icons you created in Chap. 3? This is another use for that skill! Assign each piece of text you find interesting with an icon, and abstract concepts are no barrier—you are now a seasoned sketcher!

Once you know the visual landscape of the text, think about the connections (and separations!) between different aspects of the abstract or summary. What quotes or textual summaries might you include? Mentally map out your approach to the page; this is similar to the sketchnote methodology, where you can pick a style (see Chap. 6, Fig. 6.13), but differs in that you already know how much content you want to put in.

Fig. 11.14 Makayla's response to Michael Clayton (W7) *Sketchnote LDN* Visual Abstract activity, 2018

11.6 Visual Abstracts

The sketchnote overlap also continues when you consider what else might be a visual abstract. If we think about the abstract also as a form of summary, then we can extend the practice. In Fig. 11.15 Makayla attended a panel discussion, took notes and then put the summary together after the fact, a posthumous sketchnote. This style of visual abstract is a great way of engaging others with an event and illustrating key points and is of use to the organisers. Unlike a sketchnote, it does not tell the story of the entire event and actually raises questions to provoke thought and discussion.

The more relaxed style and approach of a visual abstract allow the sketcher to take time and choose what they want to include. Whereas the sketchnote is often a fast and furious test of visual skills and iconography, the visual abstract approach means you can create a summary of just about anything. In Fig. 11.16 Makayla visited an exhibition at the Serpentine Galleries in London and created this beautiful abstract which is both a cohesive image in itself and also combines her representation of artworks from the Kamala Ibrahim Ishag, woven seamlessly into the scene.

Sketches (and thus visual abstracts) are worth a thousand words, and visual stories are more memorable than pages of written text. Consider how you might engage with this approach, and weave it into your own work and play. For example, could

Fig. 11.15 Visual summary from a panel discussion about Graphic Designers—World Anew. *Procreate* App on *Apple iPad Pro* using *Apple Pencil*. Makayla Lewis, 2020

Fig. 11.16 Sketchnote/visual abstract from Serpentine Galleries in London. Fineliner and marker on paper. Makayla Lewis, 2022

you create visual abstracts for your own research and writing and present them as part of a finished piece of work?

Practical Application Tips

- Don't try to depict too much; you have time to consider and work on a plan first!
- Define a colour palette, much as you would for a sketchnote—ensure it has good colour contrast.
- Utilise icons for complex concepts if you can, but some text can help with the overarching narrative.
- Annotate with relevant text, but don't quote huge chunks of the original work.
- You can, however, include the title in full! And add authors/artists depending on what you are creating the abstract from.
- Consider the viewer; is this going to be shared publicly? Make sure you don't include inaccurate information.

11.7 Hands-On Activities

Activity 11.1: Prompt Sketching (Group Activity)
Learning objective—Quickly sketch unexpected concepts in response to random prompts
Time—15 minutes per prompt
Materials—Fineliner, markers, coloured pencils, sketchbook or A4 paper as desired
Procedure:

- Each member of the group should write down three random phrases or concepts on a piece of paper and fold it up and put it in a box. These can be HCI related, or simply made up.
- The facilitator, or a random member of the group, should pick one folded piece of paper from the box.
- All group members then have 15 minutes to sketch their interpretation of that phrase or concept.
- After the time is up, everyone should share and describe their interpretation and sketch.
- Repeat the procedure of picking a concept and sketching twice more (or further if you are enjoying yourselves!)

This is a great way of encouraging group cohesion and discussion, and also thinking on your feet. Often, in the workplace, you'll be required to think creatively on the spot, and being able to visualise these thoughts instantly can help drive projects forward. The prompt generation can be made domain specific if you are running a workshop centred around a particular HCI or technology theme.

Activity 11.2: Sketch Data Analysis (Group Activity)
Learning objective—Learn how to perform qualitative analysis on sketched data
Time—45 minutes ×2 (two sessions)
Materials—Detailed sketches from Chaps. 6 or 7, Post-it notes (or upload sketches to *Miro* if online), black markers for writing on Post-it notes
Procedure:

- Each member of the group should bring two detailed sketches (e.g. storyboards, HCI Improv results) to the session. If it is a larger group, one detailed sketch may suffice. Sketches must all be from the same activity. Place the sketches on a large table, not overlapping and well-spaced, or stick them to a wall.
- Give each member of the group a block of Post-it notes.
- Decide as a group if you wish to allow the data to help determine the themes (inductive) or if you have preconceived ideas of what you would like to examine (deductive).
- Each group member should take 5 minutes to familiarise themselves with everyone's sketches, asking questions if necessary.

- Next, each group member should write their reactions to each image on separate Post-it notes (based on the chosen approach); stick these around the image, and when they feel they have exhausted ideas for a sketch, they should move onto the next (10 minutes).
- When everyone has exhausted their reactive ideas, the group should read and reread each other's notes and get a feel for the data (5 minutes).
- Next, the images should be moved to one side and a space made (wall or table) for all the notes to be laid out.
- As a group, with discussion, notes should then be grouped and regrouped; duplicates should be merged or linked, with the idea that they loosely form general themes. Subheadings can be used within the larger groupings (25 minutes).
- At this stage, take photographs of the groupings, and carefully pack away the Post-it notes in their groups with heading labelling the pile.
- At the next session (second 45 minutes), lay the notes back out and re-interrogate them. Have your ideas changed? Were there some notes that didn't seem to fit? Has the meaning of others been lost? If needed, revisit the sketches for clarification. Make sure the group discusses and forms consensus.
- Create a final grouping—and label your final themes and sub-themes.

This procedure can also be done on a much smaller scale, with less people, and in fact is much easier when there are less people to disagree! However, analysing data as a group and having those detailed discussions can lead to richer analysis (although we recommend no more than 6–8 people per group). Any sketched or visual data set can be used, and some researchers make their data sets open, which means you can also complete this exercise with external data and compare your findings to that of the original researchers—valuable in itself as replicability in HCI is important.

Activity 11.3: Visual Abstracts (Individual Activity)
Learning objective—Learn how to translate complex textual summaries into sketches
Time—30 minutes
Materials—Black fineliner, sketchbook or A4 paper, coloured pens or pencils as required
Procedure:

- Using *Google Scholar* or a digital library (ACM, IEEE etc.), find an interesting paper published during the last year.
- Select the abstract text; paste into the centre of a document to print or onto a digital whiteboard.
- Highlight points of interest within the text that are vital to the message of the paper.
- Sketch a connecting line to the edge of the page or to the edge of the abstract on the digital whiteboard, and sketch an icon or group of icons that represent that highlighted section. You can use short annotations.

- Look at the abstract as a whole now; is there a wider domain that this work fits into? Work out how to sketch this, as an encompassing graphic or introduction to the main focus of the paper.
- Now turn your attention to layout. You can cut out and reuse your initial rough sketches and copy paste the ones on the digital whiteboard, to help arrange your thoughts and message. Consider a fun layout style that has the right flow between aspects of the research.
- Once you are happy with the layout, create a final version of the abstract on a new piece of paper (or board area)—leaving space for title and authors! Give the finished piece a border or frame.

This can also be used as a group activity so you can compare approaches. For a wider group approach, pick papers that were all part of the same workshop or specific domain, and create a giant workshop or domain abstract on large format paper.

References

Books, Papers, and Articles

Beyer, H., & Holtzblatt, K. (1999). Contextual design. *interactions, 6*(1), 32–42. ACM.

Blevis, E., Hauser, S., & Odom, W. (2015). Sharing the hidden treasure in pictorials. *interactions, 22*(3), 32–43. ACM.

Brule, E. (2020). How to do a thematic analysis. Retrieved from Usability Geek: https://medium.com/usabilitygeek/thematic-analysis-in-hci-57edae583ca9

Buxton, B. (2010). *Sketching user experiences: Getting the design right and the right design*. Morgan kaufmann.

Clarke, V., Braun, V., & Hayfield, N. (2015). Thematic analysis. *Qualitative Psychology: A Practical Guide to Research Methods, 3*, 222–248.

Greenberg, S., Carpendale, S., Marquardt, N., & Buxton, B. (2011). *Sketching user experiences: The workbook*. Elsevier.

Harboe, G., & Huang, E. M. (2015, April). Real-world affinity diagramming practices: Bridging the paper-digital gap. In *Proceedings of the 33rd annual ACM conference on human factors in computing systems* (pp. 95–104).

Ibrahim, A. M. (2018). *Use of a visual abstract to disseminate scientific research*. University of Michigan.

Lewis, M. (2022). UX-design tools mindfulness: 'One-stop shop' for learning and practice. *Proceedings of the ICHI '22 workshop InContext: Futuring user-experience design tools*. April 29-May 5, 2022, New Orleans, LA, USA.

Lewis, M. (2023). AIxArtist: A first-person tale of interacting with artificial intelligence to escape creative block. In *1st international workshop on explainable AI for the arts (XAIxArts), ACM creativity and cognition (C&C) 2023*.

Lewis, M., Sturdee, M., Miers, J., Davis, J. U., & Hoang, T. (2022a, April). Exploring AltNarrative in HCI imagery and comics. In *CHI conference on human factors in computing systems extended abstracts* (pp. 1–13).

Lewis, M., Toselli, M., Baker, R., Rédei, J., & Ohlenschlager, C. E. (2022b, June). Portraying what is in front of you: virtual tours and online whiteboards to facilitate art practice during the COVID-19 pandemic. In *Proceedings of the 14th conference on creativity and cognition* (pp. 350–363).

Lewis, M., Sturdee, M., Gamboa, M., & Lengyel, D. (2023, April). Doodle away: An autoethnographic exploration of doodling as a strategy for self-control strength in online spaces. In *Extended abstracts of the 2023 CHI conference on human factors in computing systems* (pp. 1–13).

Sturdee, M., Alexander, J., Coulton, P., & Carpendale, S. (2018, April). Sketch & the lizard king: Supporting image inclusion in HCI publishing. In *Extended abstracts of the 2018 CHI conference on human factors in computing systems* (pp. 1–10).

Sturdee, M., Thornton, L., Wimalasiri, B., & Patil, S. (2021a, June). A visual exploration of cybersecurity concepts. In *Creativity and cognition* (pp. 1–10).

Sturdee, M., Lewis, M., Strohmayer, A., Spiel, K., Koulidou, N., Alaoui, S. F., & Urban Davis, J. (2021b, June). A plurality of practices: artistic narratives in HCI research. In *Creativity and cognition* (pp. 1–14).

Sturdee, M., Lewis, M., & Miers, J. (2022, June). Do humans dream of digital devices? Subconscious user experiences and narratives. In *Proceedings of the 14th conference on creativity and cognition* (pp. 171–183).

Websites

W1 Website of Gosh London comics – www.goshlondon.com/
W2 Website of Broken Frontier London comics – www.brokenfrontier.com
W3 Makayla's doodle gallery on Flickr – www.flickr.com/photos/makaylalewis/sets/72157648616675866/
W4 Makayla's sketchnote gallery on Flickr – www.flickr.com/photos/makaylalewis/sets/72157633090981769
W5 Makayla's personal website – www.makaylalewis.co.uk
W6 Website of the NordiCHI conference where multiple types of formats were accepted www.conferences.au.dk/nordichi2022/submit
W7 Braun and Clarke's regularly updated website on Thematic Analysis www.thematicanalysis.net

Further Reading

Agerbeck, B. (2012). *The graphic facilitator's guide: How to use your listening, thinking & drawing skills to make meaning.* loosetooth. com library.

Blijsie, J. E., Hamons, T., & Smith, R. S. (2019). *The world of visual facilitation: Unlock your power to connect people & ideas.* The Visual Connection Publishers.

Brand, W. and Koene, P. (2017). *Visual thinking: Empowering people and organisations through visual collaboration. BIS Publishers.*

Nørgaard, M. (2021). *Professional visual facilitation: A handbook for anyone working with development, facilitation, analysis or management.* Center for Visual Thinking.

Schiller, A. L. (2016). *Graphic recording: Live illustrations for meetings, conferences and workshops.* Gestalten.

Chapter 12
Sketching with Other People

12.1 Those Who Sketch Together

Sketching can also be collaborative, whether to encourage in-team cooperation and ideation or when information is needed for projects or reports. So how can we ask others to sketch for us, what might we gain from this, and how can we analyse their sketched imagery? This chapter directly follows on from Chap. 11 and covers what information or resources to provide to people, best practice and ideas for gathering, and other ways of working with sketching and people. We also provide advice for building confidence in your participants (or even your peers), as well as designing and running sketching-based user studies or workshops.

Remember how you felt starting to sketch for the first time? Around 11 chapters ago? When you ask other people to sketch for you, they will feel the same way and won't always have the time to build in confidence that completing this course will allow. So we suggest some warm-ups from the Humble Line (Chap. 2) to cover possible sketching events. The "ideas not art" mantra is vital in workshop and group situations, and we suggest you lead by example. Often we find if our sketches are demonstrated as loose and somewhat wonky (and we are perceived as "the experts") then others feel better about their own approach to sketching.

Sketching together also means working with peers, colleagues, and even continuing practice with friends. This chapter will also look at some activities that are focused on working with other people.

By the end of this chapter, you should be able to:

1. Engage participants and peers in sketching activities.
2. Learn about the value and use of participant sketches.
3. Utilise co-sketching to improve your skills and the skills of those around you.
4. Understand ethics and consent for sketching.

12.2 Why, Who, What, Where?

Why do we work with other people? Because we are human, and the human experience can never be limited to a single individual. Working with others helps us think outside of the box and our own lives and can inspire. Sketching with other people can generate insights, forge relationships, and improve communication.

Who do we work with? You could work with your peers and classmates, your friends, co-workers, teachers and lecturers, stakeholders in industry, and children; the list is endless! Sketching has the ability to break down barriers and create a level playing field. It can provide a starting off point for workshops and experiences or a method of collecting data at the end of an event. By weaving sketching into your work with others, variety and output can be enhanced.

What kind of sketches are we talking about? Everything from warming up, to generating ideas, to telling stories or designing complex technologies. Sketching as artefact, process, and output. During the course of this book, you have hopefully identified particular techniques or uses that speak to your own interests, learning, and practice.

Where might you sketch with others? If you have the right materials, you can sketch anywhere. We have conducted walking special interest groups on sketching at international conferences (Lewis et al., 2019; Sturdee et al., 2021) and hybridised versions for accessibility (Lewis et al., 2023). When the lockdown happened, we sketched virtually with people all around the world using *Miro* to share and discuss, visiting places on *Google Arts and Culture* (Lewis et al., 2022a). Most commonly, sketching with others happens at workshops or events, or during collaborative research (Sturdee et al., 2020).

12.3 Workshops and Events

The most common engagement using sketching with other people in HCI is during workshops and events. Sketching can have a fundamental role in the facilitation of workshops or be a stand-alone activity. Organising a workshop where sketching will be used is very different from taking part in generic workshops—it requires special skills and experience to execute—skills and experience which you are building right now!

12.3.1 Provide Materials!

If you have a room of reluctant sketchers, providing materials will immediately give you a head start (Figs. 12.1 and 12.2). If you expect people to bring their own pen and paper—think again. Have you ever fidgeted during a talk or meeting? Do you

12.3 Workshops and Events

Fig. 12.1 Sketching pack for students from MSc in UX module Design Thinking Theory and Practice, Kingston University London, photograph. Makayla Lewis, 2023

doodle when you are on the telephone? If you provide materials, sketching will happen… It not only levels the field in terms of materials (nobody feels left out), but it also gives people a focus for the event.

You may also wish to provide more specialist materials, depending on the type of event you envisage and if you have time constraints and want to kick-start the sketch interaction. For example, Figs. 12.3 and 12.18 show a blank and filled-in comics template into which participants can sketch and annotate their stories.

Figure 12.4 shows one of a series of five sticker sheets (images printed onto sticky-back paper) that participants could cut out and incorporate into their sketches in case they felt they could not accurately visualise a concept (Sturdee et al., 2019b). Some people only used the stickers during the event, but others placed and sketched around them. It is helpful to note that almost all participants at this conference were non-sketchers, attending a specialist computer science conference on sustainability (ICT4S, co-located with ACM Limits), yet after 4 days were confident enough to engage visually with the speakers and workshops and to display their work publicly at the end in an exhibition (Fig. 12.5). Little "nudges" are key to helping people let go of their fear of sketching!

Fig. 12.2 Participants at this conference were provided with a selection of sketching materials and sticker icons on a resource table from which they helped themselves (Sturdee et al., 2019b)

12.3.2 Get People to Engage and Relax

Nudges are also key to the major challenge of getting your participants to put down their mobile phones and laptops and actually take part (Fig. 12.6)! We do find that if you ask nicely, or get straight into the swing of things, then people tend to (guiltily) stow their digital devices. The exception might be if they are sketching digitally, but this precludes bringing sketches together with other participants, so we discourage digital sketching in real-life situations (although hybrid is another matter—see Chap. 10).

The best way to get started is to give everyone a task, a task that does not involve checking your email! For example, the portrait activity (without looking!) from Chap. 5, or perhaps getting people to sketch-introduce themselves, or even a combination approach where participants sketch and introduce each other to the group! We recommend, if you have the time, to initially start with the activities from Chap. 2, such as embracing your inner child for complete novices or for members of the public who may not know what to expect from an academic or industry workshop.

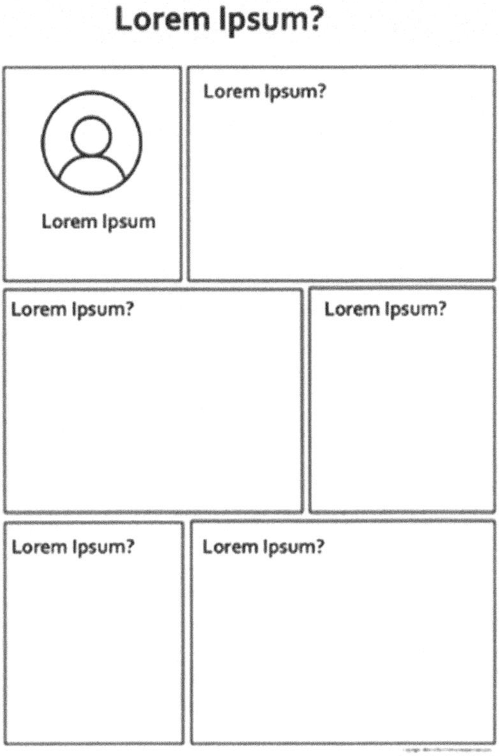

Fig. 12.3 Example blank comic template for supporting participants to sketch (Lewis & Coles-Kemp, 2014a). *Photoshop* on *Microsoft Surface Pro* using *Microsoft Surface Pen*. Makayla Lewis, 2018

12.3.3 Lead by Example and Encourage

If you find people are struggling or do not want to engage, you can help start proceedings by sketching publicly and leading by example. For certain groups, you may wish to bypass individual sketching activities and assign groups where one or two people have more confidence. If you are running an event as a collaborative endeavour, you might also assign one member of your team to each group or have them roving the room and offering assistance.

If you really need individual sketched responses, the current experience comics approach can help encourage creative interaction (Lewis & Coles-Kemp, 2014b) (Fig. 12.7). By using a previously created icon library (Fig. 12.8) (see Chap. 3 for advice), you can provide this to be cut up and integrated into an existing comics template. We discuss participant sketch outputs in more depth later in this chapter.

If you are a participant in someone else's workshop, you can still help to lead by example and engage and help your breakout group with their sketching and visuals (Fig. 12.9). Or, you may find you are helping at a workshop in that very role, and with others! If you wish to have that help, it is worth running pre-workshop sessions

Fig. 12.4 Example of printable sticker sheets to help sketch-shy participants (Sturdee et al., 2019b), based on the results of a workshop using the tactile visual library. (Lewis & Coles-Kemp, 2014a), fineliner pen on paper. Miriam Sturdee, 2019

with the people who you will be running the event with, letting them know about your expectations and any sketching skills they might be required to deploy (Fig. 12.10).

12.3 Workshops and Events

Fig. 12.5 Curating the end of conference sketch-ibition at ICT4S. Photograph, Miriam Sturdee (in Sturdee et al., 2019b)

Fig. 12.6 Sketching facilitation at a summer school at Aarhus University, lack of engagement—at the beginning! Fineliner pen and marker on paper. Miriam Sturdee, 2018

Fig. 12.7 Participant current experience comic strip depicting their use of a community centre. Fineliner pen, Post-it notes, and marker on paper (Lewis et al., 2014)

Fig. 12.8 Tactile visual library for AI readiness, *Photoshop* on *Microsoft Surface Pro* using *Microsoft Surface Pen* by Makayla Lewis for Kimbell et al. (2021)

12.3 Workshops and Events

Fig. 12.9 Makayla's sketches from ACM SIGCHI UX workshop on Futuring UX Design Tools. Fineliner pen and marker on paper and Spline. Design model. Makayla Lewis 2023 in Carter et al. (2022)

Fig. 12.10 Pre-conference sketch-facilitation workshop, during workshop, and a participant views the final exhibition. Photograph by Miriam Sturdee (in Sturdee et al., 2019b)

12.3.4 Digital Workshops and Events

Digital workshops can be an extremely valuable way of engaging with people (see also Chap. 9). Digital workshops or events are an ideal way to engage with busy people or those for whom travel is difficult or impossible. You may have sketching activities where you ask people to upload images to the board, or even draw directly onto it (Figs. 12.11 and 12.12). Key to this is leading by example, if you are also completing tasks and demonstrating your sketches, others are more likely to feel comfortable in doing so.

Fig. 12.11 *Miro* board from ACM SIGCHI conference course "The Joy of Sketch" showing the digital *Miro* board layout including title, materials, basics of using *Miro*, live demo sketch area, and a space for participants to upload their creations. *Miro* Online Whiteboard (Lewis et al., 2023)

Being able to create and curate sketches "off-screen" also helps people put pen to paper (or tablet) as they are not directly observed during the creation process. However, not everyone is used to using online digital whiteboards, so it is vital to teach people how to use them and have clear expectations and rules. For example, we provide handy hints and tips for using *Miro* in the title data of our *Miro* boards and, further, have a dedicated upload area to prevent people accidentally occluding the entire board when uploading a large sketch!

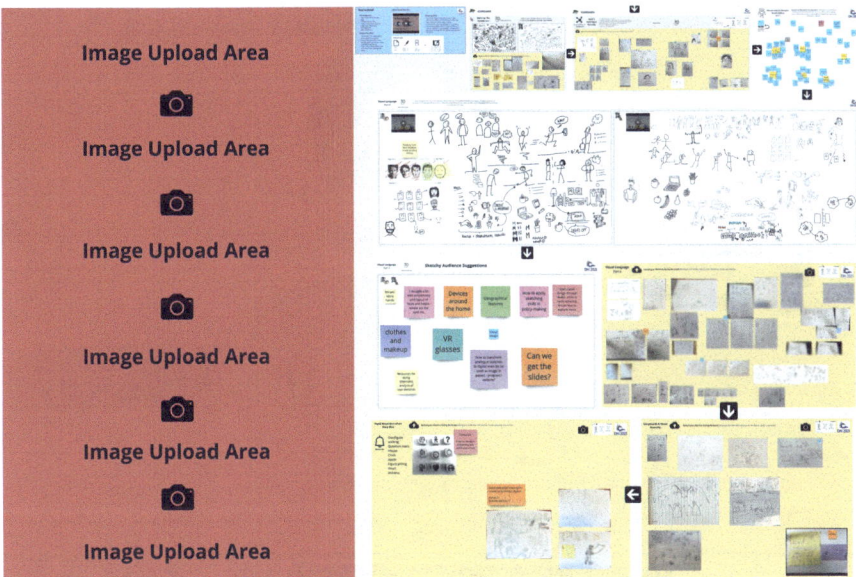

Fig. 12.12 *Miro* board from ACM SIGCHI conference course "Take a line for a walk!" showing the digital *Miro* board layout including title, materials, basics of using *Miro*, live demo sketch area, and a space for participants to upload their creations. *Miro* Online Whiteboard (Lewis et al., 2022)

12.3.5 Informal Sketching Events

These are usually attended by people who already have an interest in sketching, or learning to sketch, so you can organise your activities accordingly! For many years, Makayla ran a series of events centred around sketchnoting in and around London (Figs. 12.13 and 12.14). Whilst these have no immediate research output, the value here is in gaining experience and growing your community. Regular practice also enhances your own skills, both in practical sketching and in facilitation.

Think about whether there is a gap in the "sketching market" for a new type of collaborative event. If so, think of a focus or theme, and design activities which will appeal to a wide range of attendees. If you have an industry or academic focus, consider if you might want guest sketchers. Or could you help your attendees enhance their skills for their own research or practice? There are no limits with these informal events; be creative and engage with your community!

Fig. 12.13 Makayla displays the results of a *Sketchnote LDN* walk, talk, and sketch along a central london canal, photograph. Makayla Lewis, 2019

12.3.6 Case Study: Example Sketching in HCI Event, Workshop, or Course Structure

A clear plan is essential; share it with the participants beforehand. You should include goals, benefits, intended audience, perquisites, structure, accessibility statements, and information about yourself. Here is a summarised example from our ACM SIGCHI course, "The Joy of Sketch!" 2023 (Lewis & Sturdee, 2023).

Benefits Of all the techniques we use in our day-to-day practice for HCI and UX, sketching is the most timeless and still offers new routes into research—whether as a method, tool, or the subject of interrogation itself. Mark making is fundamentally human activity and has its roots in language formation. Sketching allows a window into the mind and offers a collaborative space to reflect and iterate. Sketching is often overlooked in many disciplines or referred to as a "soft" skill. However, it can support HCI researchers, students, and practitioners to ideate, collaborate, document, and explore and discover complex themes and spaces. This hands-on sketching course intends to celebrate and promote the diverse role of sketching to all practitioners and generate discussion—encouraging participants to adopt sketching in their everyday education, research, and practice.

12.3 Workshops and Events 229

Fig. 12.14 *Sketchnote LDN* attendees make use of a premade template to create icons for project planning, photograph. Makayla Lewis, 2018

Intended Audience The content of this course is suitable for academics (teachers and researchers), industry leaders and practitioners, students, and early career researchers who have an interest in learning and/or improving their sketching skills. Novices, experts, and those with an interest are welcome to attend.

Prerequisites There are no prerequisites, although attendees should have an interest in sketching, but prior knowledge regarding its HCI applications is not required.

Content and Practical Work The sketching in the HCI course/event will follow the following schedule:

(1) *Warm-up "The Humble Line":* Activity #1: participants will be asked to embrace their "younger selves" by mark making (scribbling); the activity aims to let go of perfection.

(2) *Icebreaker "Participant Portraits":* Activity #2: participants will be asked to work in pairs to draw each other. They will then be asked to give their drawing to the person and ask: what is your name? Where do you work, and what is your role? Why have you joined today's Sketching in HCI course?
(3) *Exemplar sketch gallery:* exemplar presentation and discussion outlining visual thinking, sketching, and sketchnotes with examples from HCI, UX, interaction design, and computer science followed by a question and answer to establish participants' key motivations and goals.
(4) *Visual language:* participants will sketch with the instructors, following a series of best practice examples that will be live drawn and digitally projected for immediacy. Activity #3: shapes, connectors, and separators. Activity #4: people, gestures, and actions + show and tell. Activity #5: scenes including buildings, places (indoors/outdoors) + show and tell. Activity #6: icon dictionary, participants will work together to rapidly build a visual dictionary of objects and concepts present in HCI. Activity #7: typography and hand lettering, participants will explore the role of annotation and notes in sketches by hand lettering using instructors' examples (worksheets) + show and tell. They were followed by Activity #8: sketchy audience suggestions, a rapid sketching session whereby participants ask the instructors to draw an action, screen, object, or concept not previously demonstrated and followed by an exemplar presentation and discussion outlining the role of colour and shading (colour theory) in sketching.

Break

(5) *Applying Sketching in HCI research and practice:* exemplar presentation and discussion outlining visual thinking and sketching from HCI, interaction design, and computer science. Activity #9: instructors and participants will produce a visual mind map exploring the significance, benefits, and pitfalls of Sketching in HCI and how they may apply it to their everyday work and research practice.
(6) *Without words:* Activity #10: participants will be asked to sketch their research area or industry practice (e.g. a recent project) without using text or verbalisation. Sketches will be placed in a "sketch gallery", an easy-to-access wall, or a large table within the course room. Using digital Post-it notes, participants will be asked to identify each sketch's field of study and critical insights; each Post-it will be stuck next to each drawing. This activity aims to get to know course peers and provide constructive feedback on narrative depiction.
(7) *Visual narratives:* storyboards and comics exemplar presentation followed by Activity #11: an instructor-led group discussion about the use of comics and scenarios in HCI, e.g. data comics, storyboards followed by best practice techniques for creating coherent and engaging comics and scenarios. Activity #12: visual economy, participants will be asked to draw a scenario/sequence in only three panels, then one meeting. This is for dissemination and publications where size/length is at a premium.

Break

(8) *Accessibility of sketches:* presentation about the accessibility of sketches in HCI, best practices, and examples, e.g. use of screen readers and the need for text alternatives (AltText) www.w3.org/WAI/alt/, and how such measures also support search engine optimisation, followed by Activity #13: participants will be asked to return to their Activity #10 outputs and add AltText followed by show and tell with a neighbour ensuring constructive critique is given.
(9) *Sketching in HCI research and practice:* design diction and speculative scenarios: exemplar presentation followed by Activity #13: group brainstorming session to explore "Applying sketching to your research/practice?". You are sketching with participants: generation and analysis exemplar presentation followed by Activity #14: instructor-led group discussion about gathering and working with participant-generated sketches.
(10) *Curricula design and pedagogy for sketching within HCI and UX education:* sketching techniques, presentation and demonstration about sharing and remote sketching sessions with colleagues, team members, and participants.
(11) *Digital sketching techniques:* exemplar gallery followed by a best practice presentation about incorporating digital sketching hardware and software in Sketching in HCI research and practice.
(12) *Resources:* a list of your favourite tools and resources.

The instructors will ensure feedback is given to each participant throughout the course. Participants will also be provided with crib sheets, practice sheets, and post-course activities (further sketching practice):

- Activity #14 Sketch Analysis: via course materials, participants will be given sketches from existing published work and taken through the methodologies that can be used to generate meaningful data from these visuals.
- Activity #15 Vignette: participants will be given a photograph (with AltText) of a *Rory's* Storycubes roll and asked to create a one-page visual story.
- Activity #16 HCI Improv: participants will be given prompts (fictional problem, user persona, technology availability, and user need) to create a visual storyboard depicting a potential solution.

Q&A, networking, then say goodbye and keep in touch!

Accessibility The course will be designed and delivered to be attended by as many people as possible. However, we will be hosting the course in person and will be happy to run a second online instance for those attending remotely. All text and verbal utterances will be clear, simple to understand, and in English. All shared imagery will contain AltText with appropriate colour contrast. All sketching demonstrations will be verbalised, straightforward to follow, and at an appropriate speed. If the course is online, remote attendees can access the conference platform through closed captioning. A virtual whiteboard platform will be set to allow for zoom-in up to 300% without problems; keyboard navigation will also be supported. Opportunities for support, questions, and comments will be present throughout the course regardless of in-person or online delivery.

Instructors Makayla Lewis is a computer science (user experience design) senior lecturer at Kingston University London, researching and teaching human factors in business, cybersecurity, smart money, and AI. Makayla is an accomplished visual thinker and sketcher who organises sketching events and courses and provides visuals for international companies and conferences. Miriam Sturdee is a lecturer at the University of St Andrews, specialising in investigating how sketching and the arts can support the design and development of novel technology. She also has an MFA in visual communication from Edinburgh College of Art.

This information has been tried and tested for Sketching in HCI courses that are in person, remote, and hybrid. You will find the activities listed here throughout the book.

Practical Application Tips

- Always provide materials!—not pencils though only pens remember sketching is plentiful.
- Select the venue right—if you want people to sketch, they need tables, not rows of chairs.
- Be very clear about what the outcome should be, and give clear instructions.
- If needed, provide an example of output first.
- Start basic; lead up to the main sketching event.
- If you see someone struggling to engage, ask them if they would like some help, and sketch a small thing to help them get started.
- Put shy people into groups with helpers, or more confident participants.
- Offer positive feedback, and celebrate all your participants' achievements.
- Help people get over their fear of sketching—lead by (wonky) example, and don't be afraid to have fun!

12.4 Graphic Recording and Visual Facilitation

Workshops are a big part of research and industry life, whether internally with colleagues or with external parties such as stakeholders and participants. We have already mentioned how sketching as a workshop participant can be useful, but you can also use your skills to help run workshops (visual facilitation) and create documentary sketches of the event (graphic recording). Figure 12.15 shows a large-scale piece which directly responded to the conversations and presentations during the event. This kind of graphic recording is especially important to help people feel "heard".

In breaks and at the end of the event, the participants came to look at the piece and could directly map their contributions to the output, and one even recognised the boat that had been used for reference (bottom left) as belonging to their friend. Having this sense of permanence and a publicly viewable record of an event is not only helpful for archiving but can become a talking point in itself.

Fig. 12.15 Visual facilitation for a "world cafe" project with University College Cork about the future and sustainability of the West Cork Islands, fineliner pen and marker on paper. Miriam Sturdee 2019

Graphic recording goes into more depth than sketchnoting, and the sketcher tries to capture as much as possible, rather than relying on subjective interpretation of events, and as such can be very time involved and requires concentration. That said, some events require only large-scale graphic recording as a more informal record, but these are useful in that they allow people who might have missed something to enjoy a "snapshot" of the talk or event and view at their leisure. For online graphic recording, you can also use digital whiteboards and curate the experience using previously created sketched resources (see Fig. 12.16).

Visual facilitation takes on a more involved role with proceedings, rather than being a passive "listener-sketcher". For example, you may be running an event and engaging other people to sketch, or you may be visualising on behalf of people in a workshop who need help working through their ideas on paper. The confidence to sketch in front of people is something that builds gradually, so don't be worried if you aren't quite ready yet. The thing to remember is they are more scared of sketching than you are.

The examples here are only a small section of the ways in which sketching has enriched our participation in and facilitation of workshops and events—for example, online events also can be enhanced with the addition of sketching—either online or offline, to be shared later (Fig. 12.17); see also Chap. 10. Additional examples of practice can be found in the Further Reading section. You may discover new ways of interaction with sketches and new exercises for participants! Engage with the activity in this chapter—and the next—for some practical experience in this domain.

Practical Application Tips

- Planning is essential for large-scale work; make time to produce preliminary sketches and translate required elements onto the large-scale format.
- For large-scale work, arrive super early if possible to prepare a border or theme, and headings—these can be neater than your "on the fly" sketches.
- When groups of people are discussing around tables or making notes, take time to quickly walk around the room, listen in (if appropriate), and see what others are drawing and writing.
- When working on large pieces, also factor in extra time at the end for extra imagery and finishing off sketches, also for chats with participants!
- Spend time curating a digital whiteboard; the overall appearance can help give a sense of professionalism, and utilising your own sketches will make it unique.
- If working digitally, make sure you "lock" your work on the whiteboard so others cannot accidentally move things around or delete them.
- On digital whiteboards, set aside an "upload area" so participants can resize and label their sketches without disrupting others.

12.5 Research with Other People

We have already covered one form of analysing sketched data for research purposes (Chap. 11), but there are other ways to elicit and gain value from participant sketches. Sometimes the very act of sketching and creative making is enough to help participants think through a problem or topic and generate data—both textual and visual (and sometimes physical, too!) (Fig. 12.17). We believe in centring sketched data, but it can also take a back seat as evidence of the process of ideation and results elicitation, perhaps as an aide to inspire dialogue or be referred back to in an interview (Fig. 12.18).

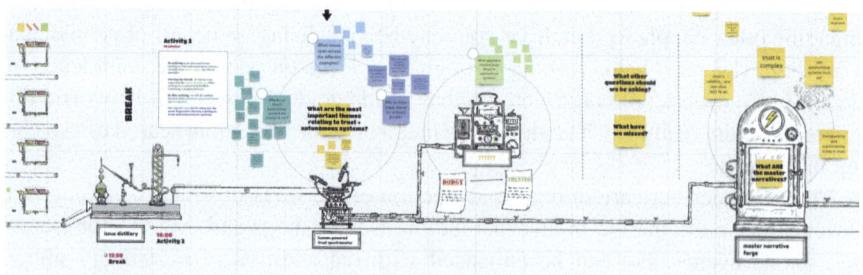

Fig. 12.16 Sketched materials to enhance the digital whiteboard experience, from TAS project. Fineliner pen and marker on paper (featured in Lindley et al., 2024)

12.5 Research with Other People

Fig. 12.17 Sketches from an online UKRI meeting about participatory research in the United Kingdom. Fineliner pen and marker on paper, Makayla Lewis, 2022

Figure 12.19 (my visual library) shows an excellent engagement using both a template, sketches, and textual input. Allowing hybrid approaches to data generation and not forcing the issue will encourage participants to experiment. Here, the participant does not feel they have to create a "perfect" outcome and have been able to iterate, delete, and rehash ideas without worry. They have also been given a safe space to operate in and felt enabled to engage with the content and topic.

In the example shown in Fig. 12.20, the research participant has sketched their idea for a tangible "kitten application" to be downloaded onto a tablet with physical properties. In the wider research study, this process was devised to understand the requirements needed to create dynamic, tangible, and shape-changing interfaces. Following the image creation (N=50), four researchers then gathered to analyse the data. Participants provided both a diagram of their imagined device and a storyboard showing its use in context. The researchers were able to deduce various

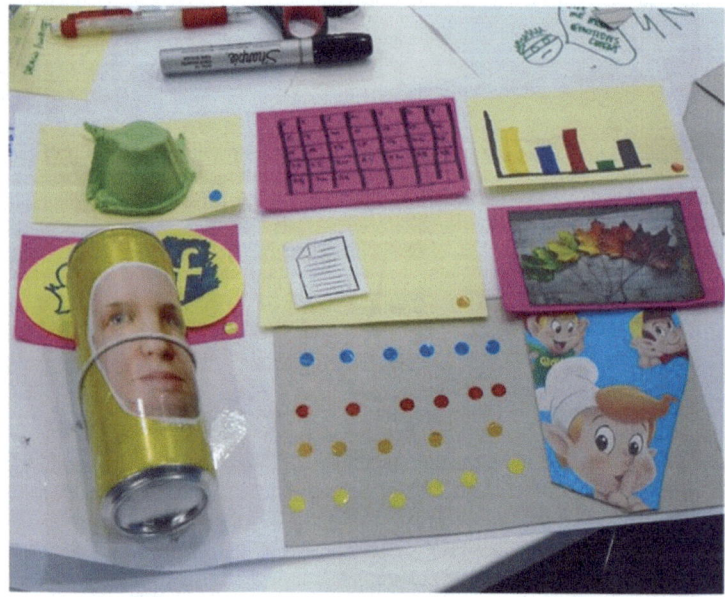

Fig. 12.18 Participants visual depiction of the future AI dashboard. Mixed media (Kimbell et al., 2021)

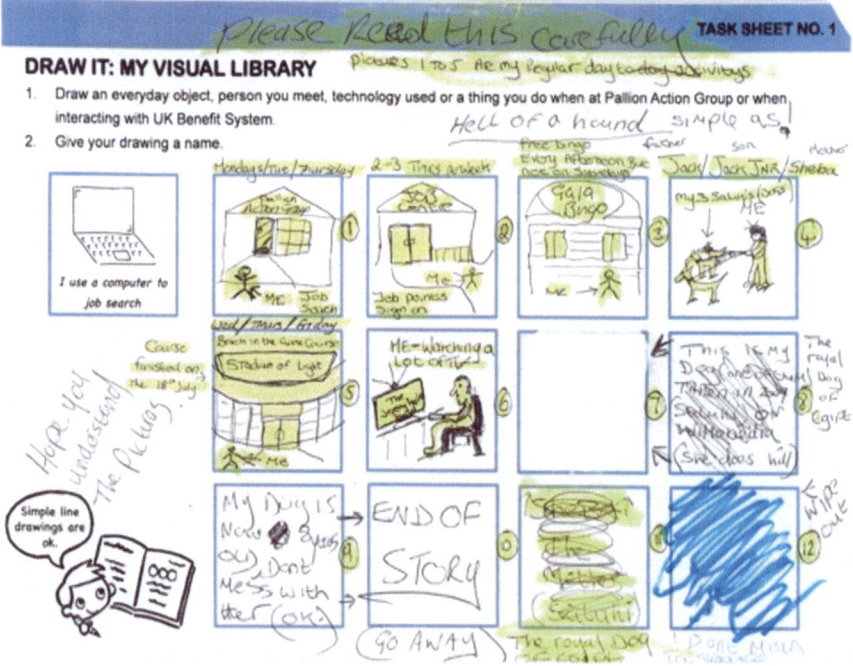

Fig. 12.19 Participant storyboard depicting their use of a community centre within a cultural probe pack. Fineliner pen and marker on paper (Lewis et al., 2016)

12.5 Research with Other People

Fig. 12.20 Participant diagram (left) and storyboard (right) depicting their idea for a shape-changing, tangible, kitten-everywhere app. Fineliner pen and marker on paper (Sturdee et al., 2019a)

requirements, such as hardware issues (weight, portability) and software (organic behaviours, lifelike interaction), and create a stack model for the future design and development of such novel interfaces (Sturdee et al., 2019a).

With more sketch-experienced people, you may find that their willingness to engage outstrips even your enthusiasm! The very first collaborative Sketching in HCI event at ACM DIS 2017 was the result in the first collaboration between us, and we engaged with sketch experts in research and practice, diving straight into the sketch-based activities and road mapping. It is testament to this first foray into Sketching in HCI that we now arrive at creating this book (Fig. 12.21).

As another example with more experienced sketchers, the large-format cocreated image in Fig. 12.22 was sketched by an arts group on an island in West Cork. Their collaborative sketching also led to collaborative verbal dialogue and ideas for encouraging sustainable practices within the island communities. At the same time, in another form of sketch-data collection, reportage sketching was carried out (Fig. 12.23)—a dual record of participation in research.

Finally, you might also collaborate with other researchers in your sketching practice, both with and without participants! Collecting multiple viewpoints and perspectives on the world is incredibly value in research and practice, to inspire and educate (Fig. 12.24).

Having a sketched conversation or being a combined visual voice during a research conversation can be a rich way of recording and making sense of information (Fig. 12.25). Bouncing ideas and imagery off another person can enhance ideation and introspection and lead to new directions for research (Figs. 12.26 and 12.27).

Fig. 12.21 SketchingDIS—exploring the role of Sketching in HCI. Fineliner pen, post-it notes and marker on paper (Lewis et al., 2017)

Fig. 12.22 Encouraging sustainable practices in the West Cork Islands. Fineliner pen and marker on paper (Sturdee et al., 2020)

Fig. 12.23 Looking inwards, sketching the sketchers fineliner pen and marker on paper (Sturdee et al., 2020)

12.6 Sketch Event Prompts

We have learnt over the past 7 years of sketching with other people, that support prompts are key to successful workshops—especially if the people you are engaging with are not regular sketchers. They are usually in analogue format but can also be provided digitally (generally using an online whiteboard); their purpose is to provide direction and reassurance, e.g. Figs. 12.28, 12.29, 12.30, 12.31 and 12.32. We recommend having a fictitious one completed that is shared with participants if they are apprehensive; this completed one should not be perfect, have mistakes, and be in a very loose sketching style. We have found that most participants believe they cannot draw; thus, the sketches they are producing are not very good. Therefore, you want to reassure them you are not asking for "art" like the masters of the eighteenth century. Think like the masters of the nineteenth century, those wonderful impressionists whose loose styles gave an idea of the world. These templates also give you a wonderful opportunity to express your creativity.

Creating blank templates with fictitious examples will encourage your participants to sketch their experiences, feelings, wants, needs, and pain points.

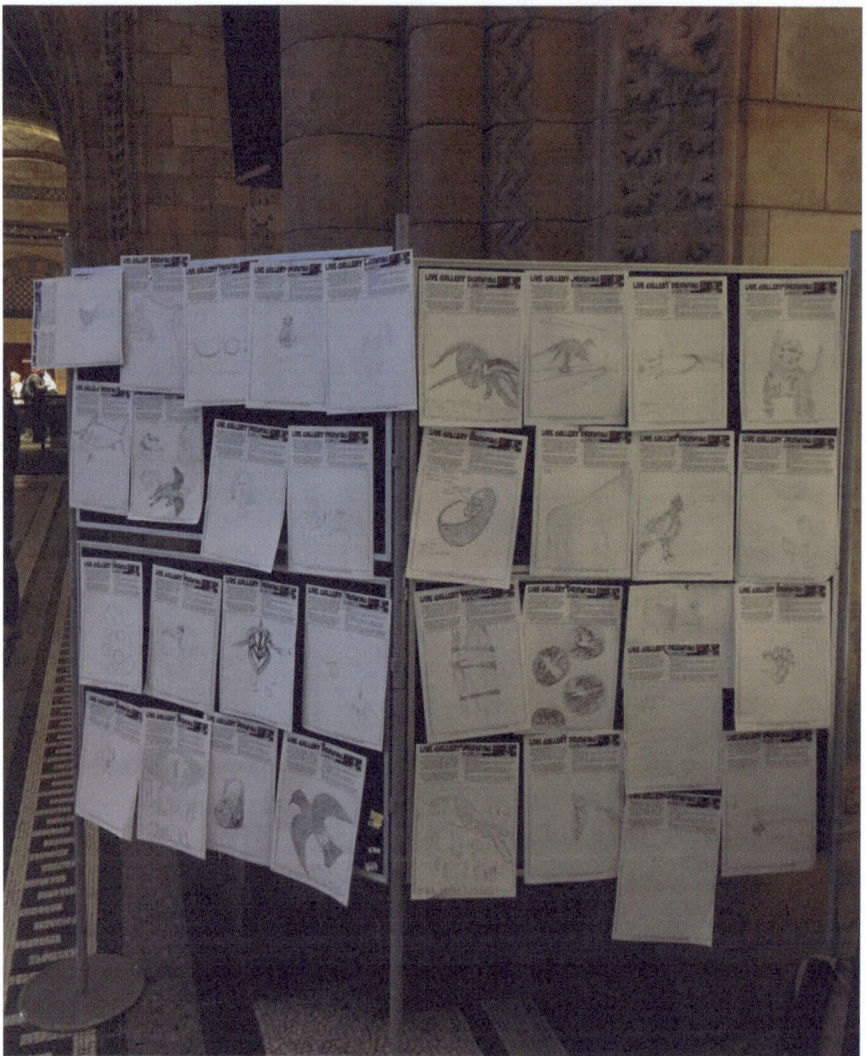

Fig. 12.24 Participant responses from Makayla's workshop at the *Natural History Museum London* for *Zooniverse*, photograph, Makayla Lewis, 2016

12.7 Sketch Ethics and Consent

Working with other people means you should engage with the ethics of good research practice. Your institution or company should follow ethical guidelines that are standard for the country you are learning or working in. For example, in universities, it is common to be asked to complete research ethics training (either in person or as an online course). Following this, there is always a process to be followed,

12.7 Sketch Ethics and Consent

Fig. 12.25 Miriam and her research collaborator sketch a road map for non-sketching participants, fineliner pen and marker on paper (Sturdee et al., 2020)

usually an ethics application to be reviewed by a panel of experts in your department or group. Some countries (or businesses) do not have formalised ethics processes embedded in their institutions, but we recommend that you follow the process of informed consent and conduct your research and events ethically regardless of this.

An ethics application will ask you about data use and privacy, and any participant materials that you may wish to use—and this means sketches! Additionally, it may mean photographs of people sketching, as you may wish to take photographs at an event to have a fast visual record of the interaction; or, if you are not confident in sketching people live, you may feel it necessary to use photographic methods as an aide memoire.

The first thing to consider is consent and/or permissions. It is common practice to ask people to sign a photo release form at an event, regardless if you intend to anonymise the images after the fact or if the results will not be published as research. If this is not needed, then you should still ask permission verbally. The same goes for using sketches by other people.

If running a research event, then you MUST get signed, informed consent (either a paper form or digital). If working with children or vulnerable adults, then a guardian or caregiver may sign the consent form on their behalf, and ONLY if the child or vulnerable person is happy with their participation. Just because sketches are not directly identifiable data does not mean they are not data that you need permission to use.

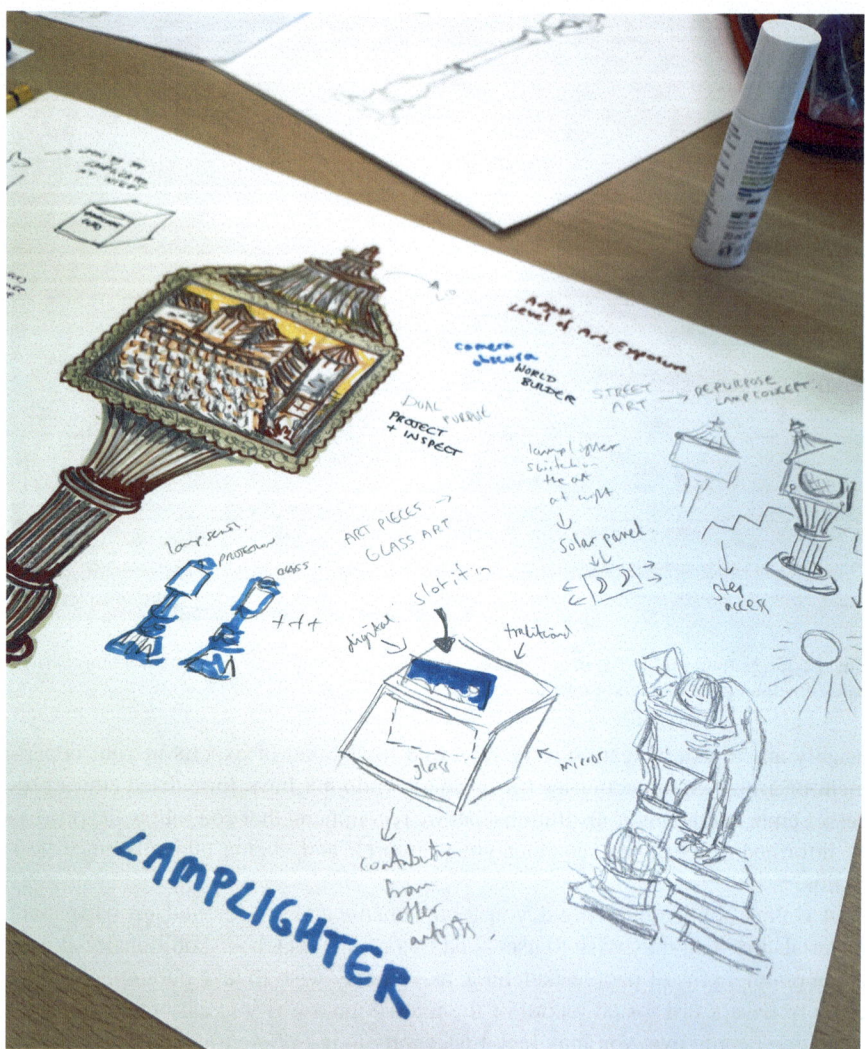

Fig. 12.26 Collaborative sketching at Venetian Drawing Conversations (Yurman et al., 2022), ACM Creativity and Cognition 2022, fineliner pen and marker on paper, Miriam Sturdee, 2022

There should be a specific consent item that allows you to reuse the sketched materials in publications, presentations, and posters (there may be other examples). When using these sketches, you must also ensure that any identifying data is removed or covered. If you are using sketches that were provided to you without these processes and were not part of the original research protocol, then you MIGHT be able to ask permission of the creator and use these with full attribution (e.g. name and date and title). In these cases the owner may also wish to assert copyright over

12.7 Sketch Ethics and Consent

Fig. 12.27 More collaborative sketching at Venetian Drawing Conversations (Yurman et al., 2022), ACM Creativity and Cognition 2022, fineliner pen and marker on paper, Miriam Sturdee, 2022

their image, and therefore you must declare this in the rights release form of your publisher.

For photographs, some people will be happy to have their image used, with the possibility of being identified, but many will not. You will see that we have already used some photographs of people in this book; however, we also have blurred or covered faces where needed (e.g. Fig. 12.33). You can also carefully compose your photographs to obscure identifying data (e.g. Fig. 12.2) Another method you can use is photo tracing (Greenberg et al., 2011); this is where you completely obscure the image of a person using outlines and basic details (Fig. 12.34). Photo tracing is a great way of presenting people actually engaging without breaching privacy and also involves some sketching! If you intend to do this, then you should make this clear to participants BEFORE you take any photographs and state in the consent that any photographs will be fully anonymised. You can either sketch over a photograph digitally or use tracing paper to create the overlays manually (Fig. 12.35).

SketCHI 5.0: Diversity & Accessibility at the core of Sketching in HCI

Sketching is a universal tool, one that has been with us from the earliest days of humanity. This freehand technique is visible both in analog and computational form using 'pencils' and 'pens', although the creation of a sketch requires human consideration and action. It is the act of sketching that we will examine in the context of cross-cultural, diverse, and accessible sketching in HCI, where it is embodied in ideation, design spaces, storytelling, impact, and much more – a sketch can be a section of code, rapid prototyping, algorithmic recognition, digital representation and more. SketCHI 5.0 will bring CHI attendees from around the world together to discuss and co-create thoughts, resources, and exemplars around the topic of Diversity and Accessibility at the core of Sketching in HCI practice.

Introduction — Please add your name, institution, research interests, and how you use sketching in your HCI research.

Sketching Tips

- **Quick and loose sketches:** The 'Ideas not art'. The purpose of SketCHI 4.0 is to discuss Sketching in HCI while practicing observation and sketching skills that can then be used during research. Please consider your sketchbook to be a piece of reportage, rather than just creating single sketch and to play with different ways of filling the space.

- **Photo-realism is not necessary:** it will not be possible, in time given, to draw everything seen. Please focus in on the details omitting or limiting backgrounds etc. to depict what is of most interest. SketCHI 4.0 is offering you an opportunity to practice the skill of 'sighting' and quick gesture capturing.

- **Build confidence:** sharing sketches can create anxiety. Consider your sketches as part of the overall 'I was there' experience.

Location ONE — **Sketch your immediate environment**
Sketch your immediate environment – i.e., mapping your real-life sketching tools including both physical and digital tools using an annotation technique. Attendees, in their groups, will be asked to introduce themselves: name, institution, research interests, and how they use sketching in their HCI research.

Makayla Lewis, Miriam Sturdee, Thuong Hoang, Mafalda Gamboa, and Pranjal Jain. 2023. SketCHI 5.0: Diversity & Accessibility at the core of Sketching in HCI. In Extended Abstracts of the 2023 CHI Conference on Human Factors in Computing Systems (CHI EA '23), April 23–28, 2023, Hamburg, Germany. ACM, New York, NY, USA, 5 pages. https://doi.org/10.1145/3544549.3583182

Fig. 12.28 Sketching two-page prompt sheet from ACM SIGCHI conference special interest group, 2023 (part 1), A4 print using *Miro* Online Whiteboard (Lewis et al., 2023)

Fig. 12.29 Sketching two-page prompt sheet from ACM SIGCHI conference special interest group, 2023 (part 2) A4 print using *Miro* Online Whiteboard (Lewis et al., 2023)

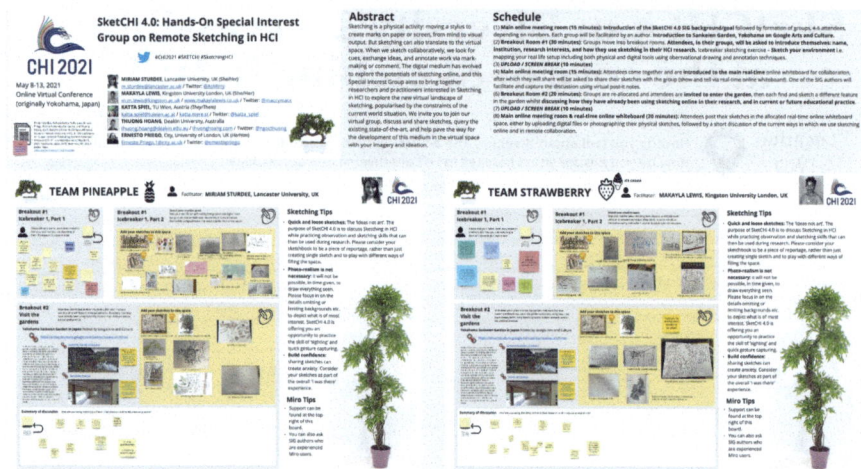

Fig. 12.30 Online sketching prompt whiteboard from ACM SIGCHI conference special interest group, *Miro* Online Whiteboard (Lewis et al., 2023)

Fig. 12.31 Practice sheet for "doodling people: garden edition" for *Sketchnote LDN*, *Photoshop* on *Microsoft Surface Pro* using *Microsoft Surface Pen*, Makayla Lewis, 2019

12.8 Case Studies

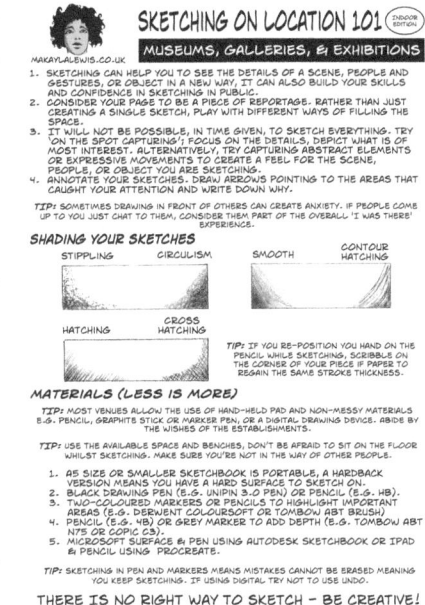

Fig. 12.32 Sketching on Location 101: museums, galleries, and exhibitions for *Sketchnote LDN*. Right side for participant sketches, left side for tips and tricks, *Photoshop* on *Microsoft Surface Pro* using *Microsoft Surface Pen*. Makayla Lewis, 2019

Please Note These are general guidelines only, and you should identify the right processes and protocols for your own situation (we cannot be held responsible for poor or unethical use of participant sketches or photographs!).

12.8 Case Studies

There follow case studies—these offer an in-depth description of some of the events we have hosted over the past 8 years for Sketching in HCI. These case studies can be adapted for your own use, and for you own events and workshops—or even large-scale conferences!

12.8.1 Case Study #1: Icebreakers and Warm-Ups

Icebreakers in workshops are hands-on activities designed to help participants loosen up (feel comfortable), get to know each other, and warm up their creative muscles before the sketching session begins. The following four exercises are our favourite:

Fig. 12.33 Online sketching prompt whiteboard from ACM SIGCHI conference workshop, *Miro* Online Whiteboard and *The Noun Project* (Sturdee et al., 2022)

Mark Marking Ask your participants to embrace their "younger selves" by mark making (scribbling) with no purpose or goal. Each participant should aim to fill a page, e.g. Fig. 12.36.

Draw with Your Nondominant Hand Ask participants to "laugh" by sketching with their nondominant hand (i.e. if they are right-handed, they sketch with their left

12.8 Case Studies

Fig. 12.34 Anonymising sketch learners at *Sketchnote LDN* meetup. The workshop handouts and example sketches were created by Makayla and give an idea of what was happening in the room, photograph, Makayla, 2016

hand and vice versa). The aim is to see perfectionism is not required and to take your sketches with a pinch of salt (as they say), e.g. Figs. 12.37 and 12.38.

Sketch Yourself Ask your participants to sketch themselves without looking in a mirror and then write a couple of sentences (name, research interest, and a fun fact); the purpose is to warm up their sketching fingers and get to know quirky things about each other, e.g. Figs. 12.39 and 12.40.

Sketch Your Creative Space (Remote Events) Ask your participants to map their real-life setup including both physical and digital tools using an annotation technique (Fig. 12.40). Meanwhile, in turns, ask them to introduce themselves (name, research interest, and a fun fact), e.g. Fig. 12.41.

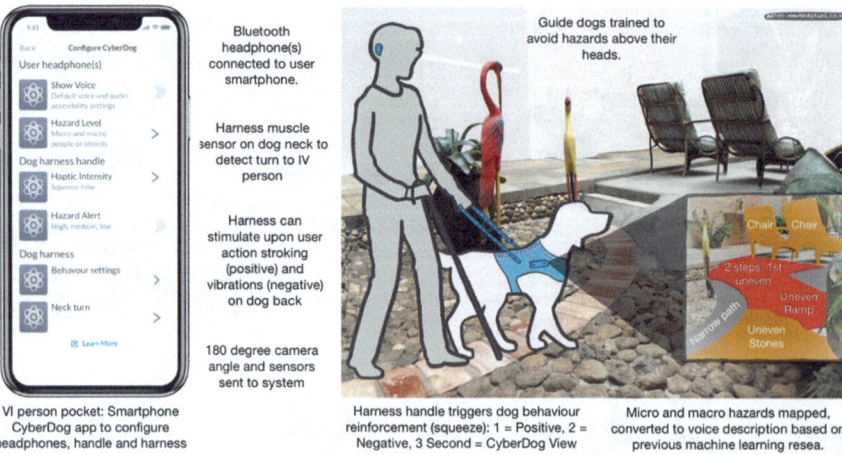

Fig. 12.35 Image tracing and sketch additions for a project about wayfinding in urban environments for people who use guide Dogs, *Photoshop* on *Microsoft Surface Pro* using *Microsoft Surface Pen*. Makayla Lewis, 2017

Fig. 12.36 Mark marking warm-up at a Sketching in HCI event, *Miro* Online Whiteboard on *Apple iPad Pro* using *Apple Pencil*. Makayla Lewis and Miriam Sturdee, 2022

12.8.2 Case Study #2: Current Experience Comic Strips (CECS)

A clear plan is essential; share it with the participants at the beginning of the workshop; confused participants equal confused outputs. You want your instructions to be clear, easy, and reasonably quick. If you would like to gather experience data that resembles a comic, we recommend the following from our collaborative work (Lewis & Coles-Kemp, 2014a, b; Lewis et al., 2022b), Sturdee et al. (2019a, b), and Kimbell et al. (2021). Here is a summarised version used in Makayla's postgraduate UX module:

Before the Workshop
- On a blank sheet of A4 paper, create 20 or more icons that represent your chosen topic. Ideally, use simple, viewer-friendly, colourless line sketches that convey these objects (e.g. Fig. 12.8).

12.8 Case Studies

Fig. 12.37 Miriam sketching a cat with their nondominant hand, fineliner pen on paper. Miriam Sturdee, 2024

Fig. 12.38 Makayla sketching a rabbit with their nondominant hand, *Procreate* App on *Apple iPad Pro* using *Apple Pencil*. Makayla Lewis, 2024

- Create a blank comic strip template; ideally, your first panel could be a space for the person to introduce themselves through a sketch and text (e.g. Figs. 12.3 and 12.7).
- Recruit two or three friends, family, or colleagues.

Fig. 12.39 "Sketch yourself" at a cybersecurity workshop, *Procreate* App on *Apple iPad Pro* using *Apple Pencil* and *Miro* Online Whiteboard. Makayla Lewis, 2023

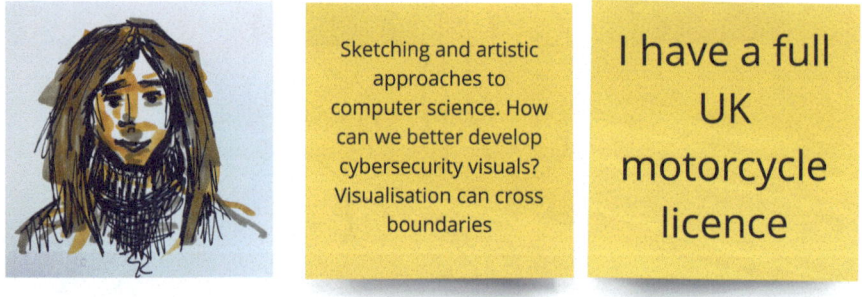

Fig. 12.40 "Sketch yourself" by Miriam at a cybersecurity workshop, fineliner pen on paper and *Miro* online whiteboard. Miriam Sturdee, 2023

12.8 Case Studies

Fig. 12.41 Sketching setup: Kitty mug full of pens, cycling glove to stop screen interference on tablet. Sketchbook page (sketched onto!), fineliner pen and marker on paper, Miriam Sturdee, 2021

During the Workshop

- Ask the participant to comic the blank CECS using the tactile, visual library, text, and their own sketches (up to 15 minutes).
- Support the participant in completing their CECS if they struggle to begin. Have a conversation with the participant as they complete their strip.
  - Refrain from being offended if the participants ask to complete the session themselves; leave and return in 15 minutes.
- Ask the participant to introduce their comic strip (5 minutes).
- Ask clarifying questions (5 minutes).
- Thank the participant for their time and provide the debrief form.

After the Workshop

- The CECS are your data, so gather them; explore the stories created, their similarities, differences, and the unexpected, and discuss (report) accordingly. You could use an affinity diagramming approach (see Chap. 11).
- Before sharing your CECS, please ensure you anonymise the data.

This information has been tried and tested for Sketching in HCI workshops that are in person and hybrid. We do not recommend doing the workshop remotely as support "on the ground" to support sketching is usually required.

12.8.3 Case Study #3: Conference-Level Sketching

Figures 12.2, 12.5, and 12.10 depict a series of events from a workshop (ACM Limits 2018) and conference (ICT4S 2018). First is a workshop designed to train and inspire people to sketchnote an event, aimed at novices with some experience in sketching. Second is a sketched summary of both the workshop and advertising the "sketching event" at the main conference. Third, conference participants were engaged using the materials provided (including icon library stickers). Fourth, created materials were displayed during and on the final day of the conference.

The main takeaways here are to identify your team—this workshop was advertised to people who had some experience in sketching and were interested in co-curating the sketching workshop experience. The main activities of the pre-conference workshop focused on establishing people's comfort levels, creating a visual icon library (tangible resource for the conference), training in "leading by example" and encouragement (using activities from this book!), and advertising the event and providing clear communication for non-sketching conference attendees.

During the conference, the workshop attendees became the sketching "champions" and worked with the other facilitators to encourage, bring materials, sketch examples, and generally provide an environment conducive to create sketching.

Before the Workshop

- Advertise to correct channels. Focus on regular conference attendees (note their engagement via social media) and members of the program committee and review team who are invested in the success of the event.
- Purchase large format paper, Post-it notes, pens and pencils, rulers, and scissors for creatively advertising the conference sketching event.
- Decide who in the workshop will be "active" sketchers (drawing alongside and with attendees) and who will be "encouragers and facilitators" (bringing materials, speaking to people).
- Co-create a tactile visual library to best represent the domain of the conference. Redraw all top-voted icons and vectors, ready to print and/or be made into stickers.
- Create posters to advertise the sketching event at the conference, A2 or bigger preferred. It can be collage.

During the Conference (Post Workshop)

- Ensure the general chairs announce the event during the opening speech.
- Have the materials table set up and ready to go, as well as evenly distributing materials across all general tables and social spaces.

12.8 Case Studies

- Set up poster boards with titles for different conference talks and events.
- Populate each event with AT LEAST one sketchnote or sketch (to be completed by all workshop attendees to ensure momentum).
- Collect sketches from those who do not have time to post them, and encourage those who have sketched but are shy.
- Clear spaces between sessions and lay out materials again.
- Curate "donated" sketches and offer to place on the poster boards.

After the Conference

- Display ALL sketches for everyone to view and enjoy—there are no bad sketches, every ones sketches are of equal value.
- Ask general chairs to signpost the exhibition.
- Clear away all unused material for recycling, reuse, and donation.
- Consider nominating the "best sketcher" at the conference for an award, to be presented at the conference closing event or dinner.

This has been tried and tested at one conference, and at smaller events. We recommend that you adapt for your particular situation and maintain regular dialogue with the general chairs. If you are planning something to be hosted at a conference and are not already on the program committee, we recommend a long lead in time as HCI conferences often are planned 2–3 years in advance. However, you can run "fringe" events without explicit organisation, though these are at your own discretion.

12.8.4 Case Study #4: Co-sketching in Research

Figures 12.42 and 12.43 show two events where co-sketching was encouraged. In the former, the sketcher (Miriam) was in charge of facilitation of a design event, both to provide sketches and reportage for the event as a whole and also to lead by example and encourage others to interact with the space, the result being a room-level installation of sketching and other creative activities. In the latter, several

Fig. 12.42 Leading by example, co-sketching at scale, Design Research Works, fineliner pen and marker on paper. Miriam Sturdee, 2022

Fig. 12.43 Leading by example, co-sketching at scale, Cork Bere Island, fineliner pen and marker on paper (Sturdee et al., 2020)

participants were encouraged to sketch, but as they were reluctant, they instead asked the two researchers to sketch items and annotate on their behalf.

The key to both events, similar and dissimilar at the same time, is to be available, listen, and create. These are both examples of very hands-on roles which require confidence, direction, and communication to ensure that the participants are getting what they need. By making the results of all image-making immediate, public, and accessible, the sketches become an important level of the conversation.

Before the Event

- Prepare a sketch kit (e.g. fineliner pen, markers, and colour pencils) that is portable and accessible and covers a range of bases. By this stage of your practice, you should have an idea of your favourite tools to work with.
- Don't expect the other organisers to provide materials… but feel free to work with whatever the world provides you with!
- Make sure there is a clear expectation of your role in the event.

During the Event

- Be prepared to ask questions! You might be asked to sketch peculiar things.
- Do not tolerate bad behaviour, language, or attitude—you are there to help not be abused.

- Say YES! (within reason). Imagine you are doing acting improv. Saying no causes blocks; give strange ideas a chance.
- Present your sketches publicly—they will inspire those around you.

After the Event

- Ensure you have a record of everything you produced.
- Take high-resolution photographs if possible, in good light, as documentary evidence.
- Summarise your role and feelings directly after finishing if possible, to capture insight, either visually or in text.
- Photograph the surroundings before you leave, to get a feel for the situation and place.

12.9 Hands-On Activities

You are nearing the end of your journey; now is the time to share what you have learned with other people! As such, there are no prescribed activities, but we suggest you review the case studies and activities in this chapter, select the ones that interest you, and mock up a sample event, workshop, or course. You should also refer back to the other chapters in the book and pick the activities you feel would work best for your particular event.

For example, you could create an interest group in Sketching in HCI in your research group, course, or department. Makayla and Miriam have co-authored many special interest groups over the years, so we recommend a hybrid approach, thus ensuring the event is as inclusive as possible. The following example is based on Lewis et al. (2023) special interest group:

Before the Event

- Goal for the event, e.g. *Diversity and Accessibility at the Core of Sketching in HCI*.
- Create a blurb for your event, e.g. *Sketching is a universal tool that has been with us since humanity's earliest days. This freehand technique is visible in analogue and computational form using "pencils" and "pens", although sketching requires human consideration and action. It is the act of sketching that we will examine in the context of cross-cultural, diverse, and accessible sketching in HCI, where it is embodied in creativity, design spaces, storytelling, impact, and much more—a sketch can be a section of code, rapid prototyping, algorithmic recognition, digital representation and more. SketCHI 5.0 will bring CHI attendees worldwide together to discuss and co-create thoughts, resources, and examples around Diversity and Accessibility at the core of Sketching in HCI practice.*
- Prepare a sketch kit that is portable and accessible and covers a range of bases. By this stage of your practice, you should know your favourite tools to work with. Don't expect the other organisers to provide materials… but feel free to work with whatever the world gives you!

- Ensure there is a clear expectation of everyone's role in the event.
- Clear and logical plan for the event: we recommend sketching the event so facilitators are on the same page, e.g. Fig. 12.44.

During the Event
- Have a fun yet related icebreaker, e.g. *draw what diversity means to you*.
- Ensure discussions are centred around an activity so the event is more engaging, for example, a sketching-related activity (discuss whilst doing), e.g. *sketching each other or the surrounding area*.
- Construct your discussion topics into straightforward questions, e.g. *How do you feel about diversity and accessibility in sketching in HCI? How do you ensure your sketches in HCI are diverse and accessible? What resources, experiences, and practices have you used or intend to use to make your sketches in HCI more inclusive and accessible? How could the importance of diversity and accessibility in sketching be better communicated in HCI and CHI?*
- Closing activity that brings together the discussions, e.g. *cocreating adaptation or iteration of the Sketching in HCI manifesto to consider inclusiveness and accessibility* (Chap. 14 for Sketching in HCI manifesto).
- Be prepared to ask questions! You might be asked to sketch peculiar things.
- Do not tolerate bad behaviour, language, or attitude—you are there to help, not be abused.
- Say YES! (within reason). Imagine you are doing acting improv. Saying no causes blocks; give strange ideas a chance.

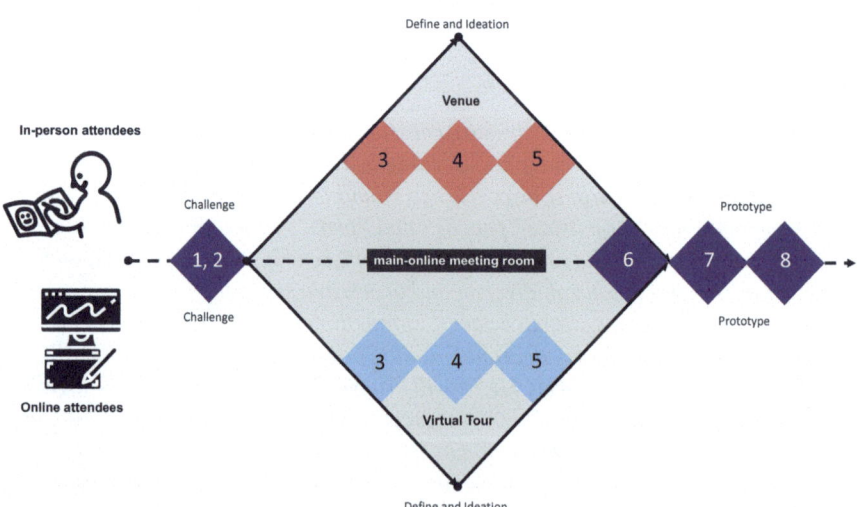

Fig. 12.44 Summary of SketCHI 5.0 schedule for in-person and online workstreams, *PowerPoint* and Diagram from Lewis et al., 2023

Fig. 12.45 Outputs from SketCHI 5.0 special interest group shared with participants, *Miro* Online Whiteboard and *The Noun Project* (Lewis et al., 2023)

After the Event

- Ensure you have a record of everything you produce, e.g. Fig. 12.45.
- Take high-resolution photographs, if possible, in good light as documentary evidence.
- Summarise your role and feelings directly after finishing to capture visual or text insight.
- Photograph the surroundings before you leave to get a feel for the situation and place.
- Send a summary to your participants, and answer any comments/questions they may have.

References

Books and Papers

Carter, A. R. L., Sturdee, M., Dix, A., Raju, D. K., Aldridge, M., Sari, E., Mackay, W., & Churchill, E. (2022, April). Incontext: Futuring user-experience design tools. In *CHI conference on human factors in computing systems extended abstracts* (pp. 1–6).

Greenberg, S., Carpendale, S., Marquardt, N., & Buxton, B. (2011). *Sketching user experiences: The workbook*. Elsevier.

Kimbell, L., Carlebach, E., Smyth-Allen, H., Gherhes, C., Lewis, M., & Vorley, T. (2021). *AI readiness: A collaborative design toolkit for professional service firms*. Oxford Brooks University/Practice Management International LLP.

Lewis, M., & Coles-Kemp, L. (2014a). A tactile visual library to support user experience storytelling. *DS 81: Proceedings of Nord. Design 2014, Espoo, Finland 27–29th August 2014* (pp. 386–395).

Lewis, M. M. & Coles-Kemp, L. (2014b). Who says personas can't dance? The use of comic strips to design information security personas. In *CHI'14 Extended Abstracts on Human Factors in Computing Systems* (pp. 2485–2490).

Lewis, M., & Sturdee, M. (2023, April). The joy of sketch: A hands-on introductory course on sketching in HCI and UX within research, practice, and education. In *Extended abstracts of the 2023 CHI conference on human factors in computing systems* (pp. 1–4).

Lewis, M., Sturdee, M., Alexander, J., Van Dijk, J., Rasmussen, M. K., & Hoang, T. (2017, June). SketchingDIS: Hand-drawn sketching in HCI. In *Proceedings of the 2017 ACM conference companion publication on designing interactive systems* (pp. 356–359).

Lewis, M., Sturdee, M., Walny, J., Marquardt, N., Hoang, T., Foster, J., & Carpendale, S. (2019, May). Sketchi 2.0: Hands-on special interest group on sketching in HCI. In *Extended abstracts of the 2019 CHI conference on human factors in computing systems* (pp. 1–5).

Lewis, M., Toselli, M., Baker, R., Rédei, J., & Ohlenschlager, C. E. (2022a, June). Portraying what is in front of you: Virtual tours and online whiteboards to facilitate art practice during the COVID-19 pandemic. In *Proceedings of the 14th Conference on Creativity and Cognition* (pp. 350–363).

Lewis, M., Sturdee, M., Miers, J., Davis, J. U., & Hoang, T. (2022b, April). Exploring AltNarrative in HCI imagery and comics. In *CHI conference on human factors in computing systems extended abstracts* (pp. 1–13).

Lewis, M., Sturdee, M., Hoang, T., Gamboa, M., & Jain, P. (2023, April). SketCHI 5.0: Diversity & accessibility at the core of sketching in HCI. In *Extended abstracts of the 2023 CHI conference on human factors in computing systems* (pp. 1–3).

Lindley, J., Benjamin, J. J., Green, D. P., McGarry, G., Pilling, F., Dudek, L., Crabtree, A., & Coulton, P. (2024). Productive Oscillation as a strategy for doing more-than-human design research. *Human Computer Interaction* (pp. 1–26).

Sturdee, M., Everitt, A., Lindley, J., Coulton, P., & Alexander, J. (2019a). Visual methods for the design of shape-changing interfaces. In *Human-computer interaction–INTERACT 2019: 17th IFIP TC 13 international conference, Paphos, Cyprus, September 2–6, 2019, Proceedings, Part III 17* (pp. 337–358). Springer.

Sturdee, M., Mann, S., & Carpendale, S. (2019b). Sketching sustainability in computing. In *Proceedings of the 2019 on creativity and cognition* (pp. 29–40).

Sturdee, M., Robinson, S., & Linehan, C. (2020, July). Research journeys: Making the invisible, visual. In *Proceedings of the 2020 ACM designing interactive systems conference* (pp. 2163–2175).

Sturdee, M., Lewis, M., Spiel, K., Priego, E., Fernández Camporro, M., & Hoang, T. (2021, May). SketCHI 4.0: Hands-on special interest group on remote sketching in HCI. In *Extended abstracts of the 2021 CHI conference on human factors in computing systems* (pp. 1–4).

Sturdee, M., Lewis, M., Gamboa, M., Hoang, T., Miers, J., Šmorgun, I., Jain, P., Strohmayer, A., Fdili Alaoui, S., & Wodtke, C. R. (2022, April). The state of the (CHI) Art. In *CHI conference on human factors in computing systems extended abstracts* (pp. 1–6).

Yurman, P., Juul Søndergaard, M. L., Pierce, J., Campo Woytuk, N., Venugopal Reddy, A., & Malpass, M. (2022, June). Venetian drawing conversations. In *Proceedings of the 14th conference on creativity and cognition* (pp. 457–461).

Websites

W1 For more of the Venetian Drawing Conversations – www.drawing-conversations-2022.com/
W2 The online book by all accepted authors from the 2022 ACM CHI workshop The State of the (CHI)Art: www.doi.org/10.6084/m9.figshare.23921814.v1

Further Reading

Barry, L. (2019). *Making comics*. Drawn and Quarterly Illustrated Edition.
Baskinger, M. (2013). *Drawing ideas: A hand-drawn approach for better design*. Watson-Guptill Publications Inc. Illustrated edition.
Gray, D., Brown, S., & Macanufo, J. (2010). *Gamestorming: A playbook for innovators, rule-breakers, and changemakers*. O'Reilly Media, Inc.
Hoffmann, A. R. (2019). *Sketching as design thinking*. Routledge.
Scobie, L. (2018). *365 days of drawing: Sketch and paint your way through the creative year*. Hardie Grant Books (UK).
Sibbet, D. (2010). *Visual meetings: How graphics, sticky notes and idea mapping can transform group productivity*. Wiley.
Smith, K. (2013). *Wreck this journal: To create is to destroy, now with even more ways to Wreck!* Penguin.

Chapter 13
The Future of Sketching

13.1 Introduction

Whereas Chap. 1 described the history of Sketching in HCI and how we might apply sketching within its context, here, we explore the future and what the sketch and sketching practice might become over time. Sketching has long been a valuable process in design, engineering, and science, as well as the arts and humanities—but how does this traditional, often ephemeral practice fit into the futuristic world of HCI? Recent advances in computing, AI, and robotics are changing the face of image creation. What other areas might come to the fore in the near or far future?

The breadth and depth of how the sketch has been adopted by HCI and within computer science over the past 60 years are an example of its interdisciplinary potential, and possibilities for use in technical contexts. Back in 2002, some suggested that traditional sketching might be under threat from computer-aided drawing (Jonson, 2002)—but this is not the case: more than ever before, people are taking up pen or stylus to record visual ideas, and courses and workshops in sketching are a regular occurrence in conferences and other events.

This is also only one part of the sketching HCI revolution; the sketch is taking on new life in the hybrid forms and as an intermediary in interfaces and workflows. Defining and documenting the sketch in HCI are an interesting focus, but the real challenge comes in the teaching of freehand sketching and drawing as part of the university curriculum, or even sooner, in our schools, where the arts frequently face cutbacks in funding. Students are often reluctant to learn to draw unless they are "a natural" or put off trying by those around them who appear to have more talent for it (Cohn, 2012). Johnson et al. (2009) argue that the "sketch" in computing is a niche area, but we disagree; a sketch is more than that; it encompasses many domains and fields, and the computerisation of sketching and its adoption by HCI will only expand its reach.

Sketching is commonplace within teaching practice for architecture, design, and engineering but has not previously been widely adopted in HCI and computing. With the advent of "beautification" programs, sketch-assist software, and gamified drawing programs (Williford et al., 2017), this could signal overall change. However, there is an elephant in the room—the rise of machines, specifically artificial intelligent agents such as *Midjourney* (W1) and *ChatGPT* (W2).

We provide a position piece and speculative ideas about sketching in computer science, research, and industry and how sketching practice might evolve. We also invite students to consider their own futures and how they will use sketching in their professional or future research and learning practice.

Have we reached "peak" sketch in HCI? Or are we about to embark on a renaissance, embracing hybrid forms and interdisciplinary collaborations?

By the end of this chapter, you should be able to:

1. To fight or hug the rise of the machines?
2. Learn to value your individual and human creativity.
3. Learn to use LLM agents as a source of creativity, not a replacement.

13.2 Teaching and Learning

Sketching is commonplace within teaching practice for architecture, design, and engineering but was previously less widely adopted in HCI and computer science. With the advent of CAD, *Photoshop*, "beautification" programs, sketch-assist software, and gamified drawing programs, HCI can take centre stage in sketching education, which could be a focus area in coming years, e.g. Figs. 13.1 and 13.2 Miriam and Makayla sketching in technology and HCI settings.

Sketching is pivotal in helping researchers' practitioners communicate their thoughts, ideas, understanding, knowledge, and outputs. The rise of digital sketches and AI does not mean there is a lack of exciting research for analogue sketching (and this book illustrates this fact!), merely with the quantity of publications, as "traditional" sketching may have become more incremental in its advancements—but research in this area is inexorably moving forward. One explanation for the possible lack of focus on sketching may be that the idea of *what* a sketch is has changed (although Goldschmidt states that the sketch (traditional view) is still very much relevant (Goldschmidt, 2017)).

We don't have all the answers yet, only to state that traditional and analogue or digital hand sketching is definitely still relevant, appreciated, and needed—there has been a rise in recent years in sketching courses in HCI, computing, and UX and renewed interest in its application in HCI via workshops and tutorials at high-profile venues.

The current methods of teaching sketching and creativity will never go away, but we now need to think of how we might design courses which incorporate the new technology, and the future of technology? Watch this space!

Fig. 13.1 Makayla teaching sketching at AfroTech Fest in London. Fineliner pen on Post-it notes and *Procreate* App on *Apple iPad Pro* using *Apple Pencil*. Makayla Lewis, 2018

13.3 Possibilities for Sketching with Emerging Technologies

At the beginning of the twenty-first century, a leap in technology innovation and integration heralded the digital age. Today's technologies have undergone significant and rapid advancement, allowing us to think, create, and communicate in various exciting ways. Sketching has been integrated into physical interfaces so that blind and visual impaired people can sketch with haptics and into drawn sculptural images by combining 3D printing with drawing (*3Doodler*—W3). Whilst advances in freehand sketching have been made, the notion of magic paper—seamless computational sketching—is still some way from being fully realised. However, some drawing tablets have come close to paper texture with matte surfaces and flexible OLED (organic light-emitting diodes, a type of screen hardware which is bendable). Additionally, there are other forms of tangible interface that are becoming more

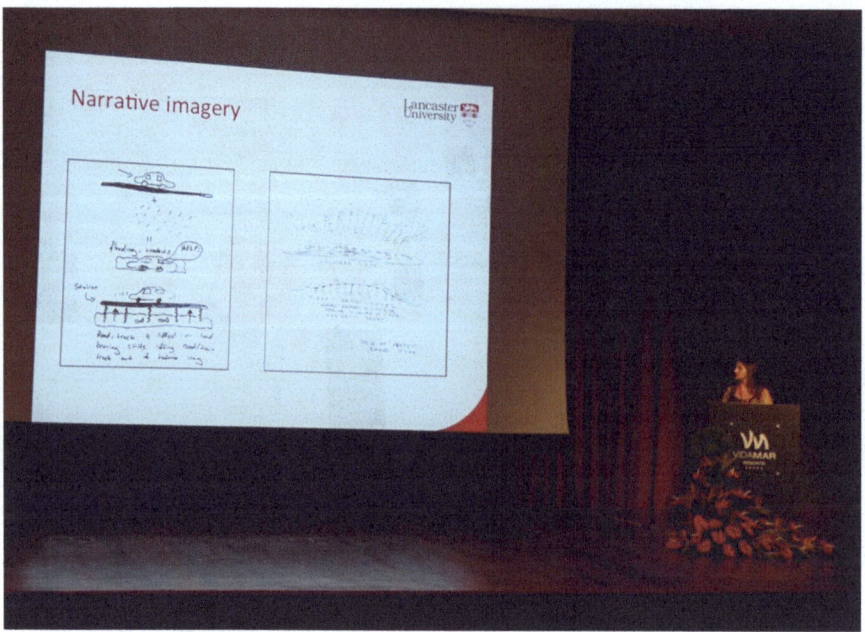

Fig. 13.2 Miriam talking about sketching and narrative imagery for shape-changing interfaces at ACM ITS conference 2015. Fineliner pen on paper (Sturdee et al., 2015)

sophisticated, and newer versions of sketching technology are becoming commercially available.

The world of technology, creativity and sketching is ever-evolving, and the needs of sketchers and viewers' expectations will remain a fundamental skill and practice embedded in our work, research, teaching, and learning. Therefore, the future of Sketching in HCI is enticing; the possibilities could reshape the tools to draw, the process of creating a sketch, and how we interact with sketches. This book will encourage further pursuit of these possibilities.

The march of innovation will not ignore manual sketching! We imagine that, in the year 2030, we *could* see greater or new explorations of…

- **Digital sketching tools**—More sophisticated, accessible, precise, and versatile sketching software, apps, platforms, and analogue mediums, thus broadening the audience of sketchers and viewers and their abilities and efficiency to create.
- **Cross-platform and medium integration**—Integration of sketching tools across various platforms and devices could allow sketchers to ideate and craft an analogue sketch; continue a digital gadget on the go, then in a static location; return to analogue; hand over to a collaborator; and then conclude on another device quickly, easily, and effortlessly.
- **Augmented reality (AR), virtual reality (VR), and mixed reality (XR)**— Immersive sketching in 3D space could allow sketchers to work in ways impossible in the physical world. AR could enable sketchers to overlay digital images

onto physical spaces; this could support design fiction and speculative design research and practice.
- **Real-rime collective sketching**—Multiple sketchers could create in-person, remotely, and hybrid using numerous software, apps, and platforms simultaneously. This could be useful for supporting brainstorming, creativity, and annotation of created sketches.
- **Internet of Things with VR, AR, and AI**—Intermingling AR and VR with IoT devices with integrated AI could support sketchers to create intelligent, context-aware, and interactive sketch environments that respond to their (and reviewers) needs, preferences, creations, and skill levels.
- **Tangible and shape-changing interfaces**—Bridging the gap between digital and physical sketching could allow sketchers to interact with physical objects to draw or manipulate digital sketches in real time individually or collaboratively, which could be used for design ideation and prototyping. We might also explore how we sketch on surfaces that are unpredictable and have their own agency.
- **Robotics**—There has been considerable interest in robot sketching, be it using drawing machines with human input (like the *AxiDraw* (W4)), or the creation of artist-robot *Aida* (Romic, 2022) or *Paul the Robot* (Tresset & Leymarie, 2013) who is designed to "sketch people". The problem with robotic creativity is that it is fed by raw data and human input—is this real creativity? What next?
- **Natural user interfaces**—Support sketchers to create and manipulate sketches using their voice, reducing the need for traditional input devices.
- **Neural interface**—Brain-computer interfaces (e.g. Shafiei & Esfahani, 2014) could play a role in sketch creation.

These are only some ideas we have begun to see or hear about; there are probably many others. Do not let this list hinder your future thinking and sketching. Please do share your findings (and practices) with us, we are always interested in what others are up to, e.g. (Lewis, et al., 2024).

13.4 Possibilities for Generative AI

The future of sketching and artificial intelligence (AI) is promising as technology advances. Whilst it is uncertain how far the integration will go, arts and AI will likely influence one another; generative AI-powered sketching tools, e.g. *ChatGPT* (W2), *Midjourney* (W1), and *Open AI DALL-E* (W5), could provide auto-correction and context-aware suggestions and even generate sketches based on user input, thus assisting and enhancing sketchers' creative process by developing multiple design options.

13.4.1 Generative AI and Sketching

To explore these possibilities further, we interacted with *ChatGPT*, prompt: **How can *ChatGPT* help sketchers in HCI to become better at learning and integrating sketching into their everyday work, teaching, and learning?**

"ChatGPT can assist sketchers in Human-Computer Interaction (HCI) in various ways to improve their learning and integration of sketching into their work, teaching, and learning. Here are some ways in which ChatGPT can be helpful:

1. **Tutorials and Guidance:** ChatGPT can provide step-by-step tutorials and guidance on sketching techniques, tools, and best practices in HCI. It can answer specific questions about sketching methods and offer tips to improve sketching skills.
2. **Feedback and Critique:** Sketchers can share their sketches with ChatGPT and receive constructive feedback and critique. This can help them identify areas for improvement and refine their sketching abilities.
3. **Idea Generation:** ChatGPT can assist in generating ideas for sketches, design concepts, and user interfaces. It can provide creative prompts and brainstorming assistance to inspire new sketching projects.
4. **Prototyping Assistance:** When sketchers are working on HCI prototypes, ChatGPT can help them by suggesting ways to translate sketches into digital mockups or wireframes. It can recommend relevant tools and techniques for prototyping.
5. **Design Patterns and Principles:** ChatGPT can provide explanations and examples of HCI design patterns and principles to help sketchers create more effective and user-friendly interfaces.
6. **Case Studies:** It can share real-world examples and case studies of successful HCI sketching projects, highlighting the benefits of integrating sketching into the design process.
7. **Teaching Resources:** For educators and trainers, ChatGPT can offer course materials, lesson plans, and resources for teaching sketching in HCI. It can provide suggestions for classroom activities and assignments.
8. **Research Assistance:** Researchers in HCI can use ChatGPT to access academic papers, research findings, and the latest trends in the field related to sketching and user interface design.
9. **Collaborative Work:** ChatGPT can act as a collaborative partner, offering ideas and suggestions during sketching sessions. It can facilitate brainstorming sessions and contribute to the ideation process.
10. **Troubleshooting:** When sketchers encounter challenges or roadblocks, they can turn to ChatGPT for troubleshooting advice and solutions to common issues in sketching and design.
11. **Community Building:** ChatGPT can connect sketchers with online communities and forums where they can share their work, seek feedback, and collaborate with others in the field.
12. **Motivation and Inspiration:** ChatGPT can provide motivation and inspiration by sharing success stories, quotes, and motivational content related to sketching and HCI.

Overall, ChatGPT can serve as a versatile and knowledgeable assistant, offering both practical advice and theoretical insights to help sketchers in HCI enhance their skills and seamlessly integrate sketching into their everyday work, teaching, and learning".

We believe that points 3, 4, 5, 9, and 12 will be possible (some are occurring now; see following sections), whereas points 1, 7, and 10 could be possible yet superficial. Although points 2, 6, 8, and 11 will be lies… in fact the results we elicited here could also be seen as applicable to many different subjects!

Whilst interesting and helpful, LLMs (large language models) like *ChatGPT* or *Midjourney* cannot help you think through problems in your own head, and they cannot be independently creative—they NEED us. An LLM cannot really provide meaningful HUMAN-sketching interaction advice, and further, the image-based LLMs can provide "sketch effects" on existing work they have been trained on but will never replace the human sketcher, as the process behind the design of images cannot be forced into a simple digital question. By trying to offload this sort of task to an agent, we miss out on the most valuable aspect of ideation—novelty and creativity.

If you have got this far with this book, then you don't need to ask *ChatGPT* for help (except perhaps as a sketching prompter!). *ChatGPT*, *Midjourney*, *DALL-E*, and other LLMs should be tools to assist people, not replacements for actually sketching and making. That said, it does encourage many people who are not artists or sketchers to try to create beautiful imagery, but all that is generated is based on real work by illustrators and artists around the world, so it is homage, rather than individual expression…

13.4.2 Sketching with Generative AI as a Collaborator

Generative AI could become a collaborator or an assistant for sketchers, support creativity, provide inspiration and directions, and provide overarching creativity support. Early engagement with generative AI has led to growing concerns amongst sketchers (and more comprehensive creative practitioners and educators) regarding its ethical use in the sketching and creative process. In a recent first-person workshop paper, Makayla Lewis (2023) shares their experiences of interacting with generative AI to escape creative (art) block (Figs. 13.3 and 13.4) and results (thoughts and feelings) of such interaction (W6):

"*ChatGPT* and *MidJourney* represented the missing art teacher who responded to my questions and reiterated/explained misinterpretations—a guide, as you will. The artwork created was inspired by the AI responses, Makayla Lewis: 'I do not see this work as personal but as guided mark-making to escape a creative block. Thus, it would not be used or presented as commercial work'".

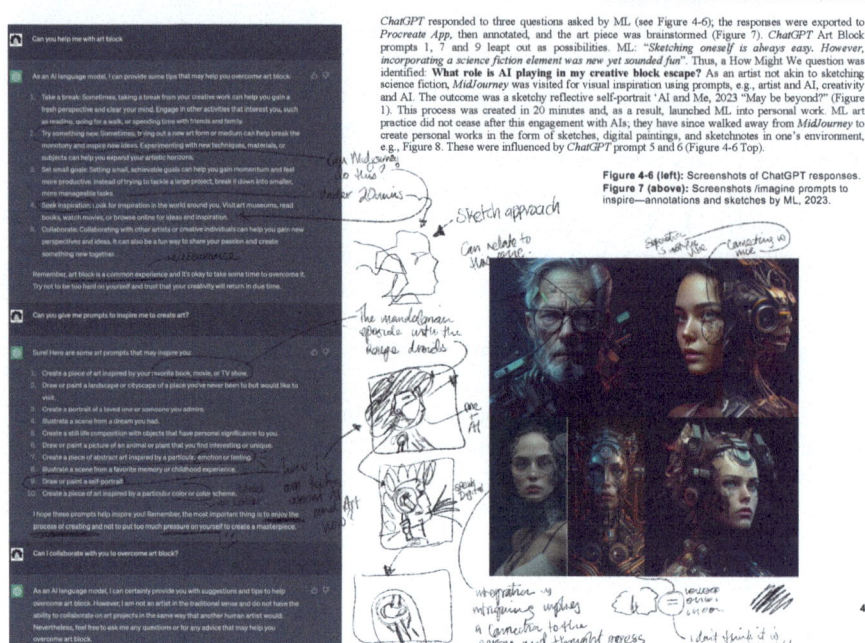

Fig. 13.3 Extract from Makayla's workshop position pictorial exploring the use of *ChatGPT* and *Midjourney* to support removal of creative block. *Procreate* App on *Apple iPad* Pro using *Apple Pencil* (Lewis, 2023)

As a result, they identified the following observations that require greater consideration:

- **Transparency of attribution**—AI views contributions of artists across the Internet (studying the work of different artists), but who are the contributors? Why were they selected? Were permissions obtained? And does this lack of transparency impact trust in the responses or negatively impact the arts community?
- **The creation process**—Studying others (human or AI) requires a clear understanding of the creation process; how and why were responses constructed in such a way?
- **Ethics of asking**—Is a quick fix, nonhuman engagement, to escape creative block appropriate? Would engagement with the physical world be more beneficial to boosting creativity?
- **Inspiration vs. copying**—Art practice and personal art production have different roles in artists' creative process; ensuring they remain separate is vital to a thriving arts community. Thus, how will AI's art prompts and visual generators ensure separation?

There is a possibility for generative AI to support sketchers like yourselves to remain creative and good at sketching. However, greater exploration of explainability and ethical responsibilities that occur because of using them still requires

13.4 Possibilities for Generative AI

Fig. 13.4 AI and Me, 2023 "May be beyond?". Digital sketchbook entry, a response to *ChatGPT* prompts and *Midjourney*/imagine inspirational outputs. *Procreate* App on *Apple iPad Pro* using *Apple Pencil* (Lewis, 2023)

research. Thus, we recommend that if you use (or intend to use) generative AI to support your engagement with Sketching in HCI, that you are mindful of the above considerations and the current trends in research and practice. We do not encourage or discourage its use; when writing this book, AI is still emerging and finding its feet within sketching (and the broader arts). However, we ask you to be responsible and informed when interacting with such powerful technologies.

Practical Application Tips from *ChatGPT* (Fig. 13.5)

If you want to explore the future via sketching with LLMs, start with Hands-On Activity 13.2.

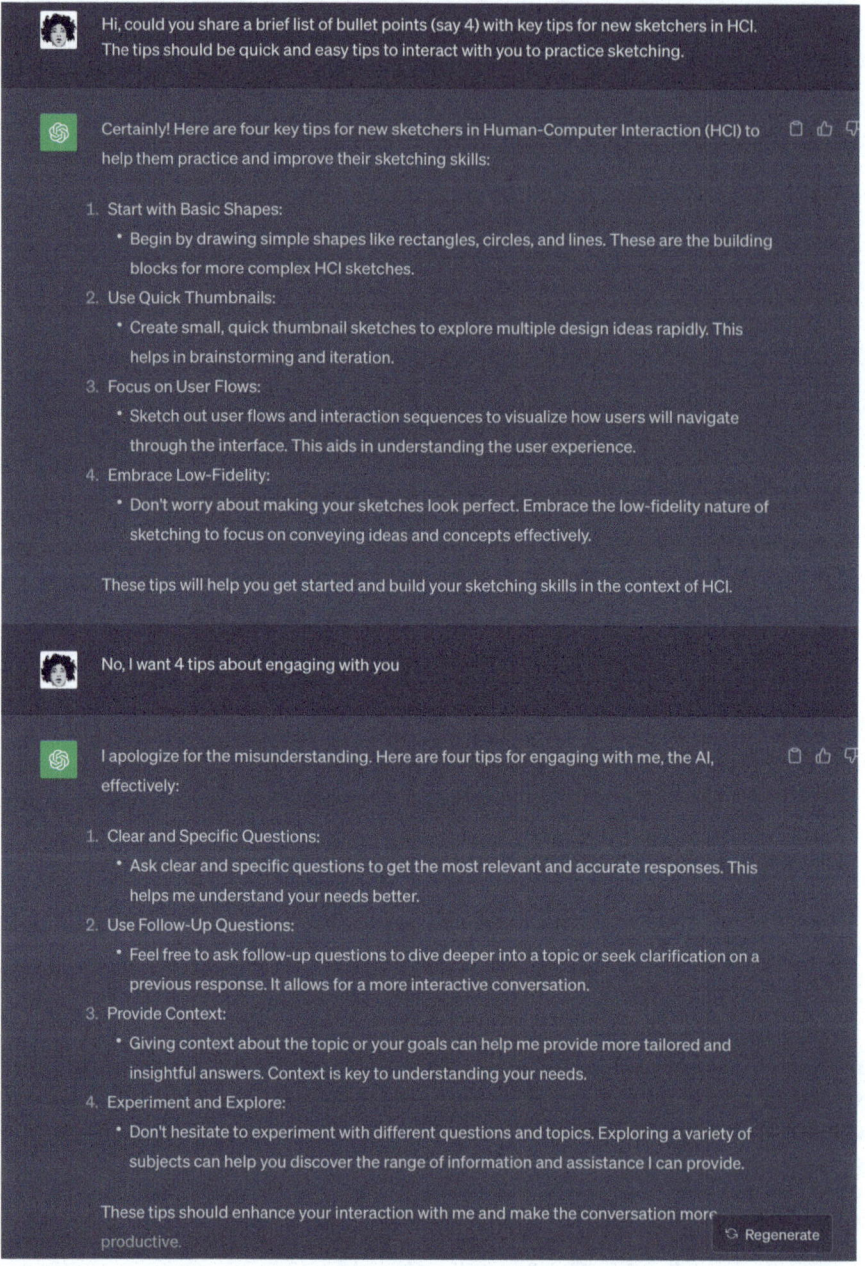

Fig. 13.5 *ChatGPT* responses to prompts: "Hi, could you share a brief list of bullet points (say 4) with key tips for new sketchers in HCI. The tips should be quick and easy tips to interact with you to practise sketching" and "No, I want 4 tips about engaging with you". *Procreate* App on *Apple iPad Pro* using *Apple Pencil* (Lewis, 2023)

13.5 Sketch and Sustainability

It would be irresponsible to end this chapter without considering climate change and sustainability. Every facet of human-computer interaction should consider how our general waste, e-waste, and carbon footprint have an impact on the world and lead to climate change (Thomas et al., 2015). We are quick to upgrade, change, and break the technologies we are gifted with (Sturdee et al., 2020) by the continual march of innovation and never-ending hype cycles.

Sketching digitally is subject to the same forces of decay as any other technology used to enhance the human experience, and sadly, sketching using paper and pen also produces waste—human beings create endless ephemera.

The advantage of digital sketching is that we do not create endless paper waste, but each image file created and its processing use power which in turn impacts the world we live in.

The advantage of analogue sketching is that we can recycle paper and sketch with and on almost anything. From the back of the napkin to scratching with a stick on the beach, improvising is possible. However, we use and consume very real resources such as trees and carbon and plastic for our pens.

There is no ideal way to be sustainable, but we can try our best to prevent excessive waste and usage of resources. In the future, someone may find a technological solution. Could it be you?

Practical Application Tips

- Reuse scrap paper whenever you can, e.g. the back of junk mail, notes from meetings, envelopes, etc.
- Buy refillable pens and markers rather than one-use disposables.
- Do not upgrade your digital drawing tablets because a shiny new product has been released, unless it will dramatically improve your sketching experience. If you choose to upgrade, ensure you sell your old drawing tablet on the second-hand market, send it to recycle programs, or gift it to a friend, family member, or a charity.
- As sketchers in HCI, we love good Post-it notes (so do Miriam and Makayla); we implore you to use them sparingly; you could switch to online whiteboards such as *Miro* to collaborative sketch on Post-it notes. If you prefer the tactlessness of paper, please ensure you recycle when the Post-it is no longer needed.
- Every sketcher should have access to a whiteboard or blackboard in their creative space (indoors and outdoors) because it provides endless sketching with little waste. Makayla recommends, e.g. an A4 whiteboard notebook for on the go (e.g. (W7)) and A3 whiteboard static sheets for indoor walls (e.g. (W8)) with refillable dry-erase markers (W9).

13.6 Hands-On Activities

Activity 13.1: Sketching Timelines (Individual Activity)
Learning objective—Develop a sketch-based timeline of your future practice and milestones
Time—10 minutes–?
Materials—A blank piece of paper (A4 or A3) or a page/spread of a sketchbook (although make sure it is at least A4), any pen or pencil available, coloured pens/pencils if desired. You may also wish to do this digitally, but having a physical reminder of progress can be beneficial.
Procedure:

- Choose either a horizontal or vertical layout.
- Draw a line down the middle of the page; at one end put today's date, or the month and year.
- Mark out a time period of your choosing by putting a small line through the main divide for each segment, leaving equal gaps between each.
- Decide upon your end goal—is it to become better at observational drawing? Is it to incorporate sketching into a consistent everyday practice or to utilise it for a particular project? Write the goal at the top of the timeline.
- Imagine what steps you will have to take to achieve this goal, and put a dot along the timeline where you think these steps are practical.
- On another page or piece of paper, design an icon for each step; once you are happy, draw them along the timeline near their corresponding dot—making sure to swap sides so each one has plenty of space around it.
- Draw a circle or other shape around each icon, and connect each to its dot on the timeline; label or number them accordingly (e.g. name, date, or sequence).
- As you reach each milestone, draw a constellation of smaller icons, figures, or images around it to represent the activities you took, and link these up. Use colour, detail, and annotation if you wish. The goal is to create a documentary piece of your journey into sketching following completion of this beginners' course!
- You can also use this to document a thesis, project, or any other goal of your choosing. If you choose to use a piece of paper, you can make a poster to remind you of your progress.

Activity 13.2: *ChatGPT* and Friends
Learning objective—Interact with generative AI to practice sketching
Time—30 minute
Materials—A blank piece of paper (A4 or A3) or a page/spread of a sketchbook (although make sure it is at least A4), any pen or pencil available, and coloured pens/pencils if desired. You may also wish to do this digitally, but having a physical reminder of progress can be beneficial. You will need access to the Internet to complete this activity.

13.6 Hands-On Activities

Procedure:

- Go to www.chat.openai.com/ (you may need to create an account to use the platform).
- Ask *ChatGPT*, "Hi, I hope you are having a good day. I want to sketch, but I do not know what to draw. Can you give me five one-line prompts that I can sketch in 5 minutes?"
  - Once you can sketch the first prompt, ask the question again.
  - You will get different responses every time you ask the above question, e.g. Fig. 13.6.
- Now, write your prompt and sketch your response.
- Now go to www.midjourney.com/ (you make need to create an account to use the platform).

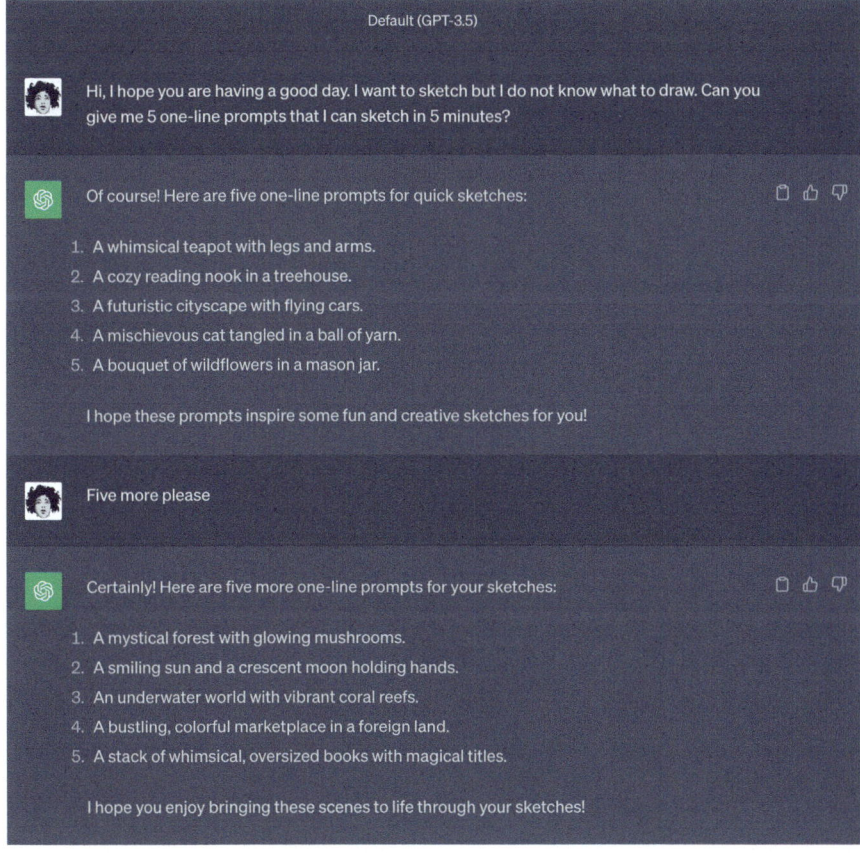

Fig. 13.6 Exemplar from *ChatGPT* "computer science and sketching". *Procreate* App on *Apple iPad Pro* using *Apple Pencil* (Lewis, 2023)

- Ask *Midjourney* to "/imagine sketching and computer science".
  - You will get different responses every time you ask the above question.
  - Ask this question four times so you have four images (e.g. Fig. 13.7).
  - Using Midjourney gothic style inspirational visual responses, create a sketch that depicts sketching in human-computer interaction. Let the reactions inspire your output.
- Finally, go to another generative AI of your choice, e.g. www.openai.com/dall-e-3. Create a prompt, ask the prompt, and let the responses inspire another sketch.

Fig. 13.7 Exemplar from Midjourney "computer science and sketching". (Lewis, 2023)

References

Books and Papers

Cohn, N. (2012). Explaining 'I can't draw': Parallels between the structure and development of language and drawing. *Human Development, 55*(4), 167–192.

Goldschmidt, G. (2017). Manual sketching: Why is it still relevant? In *The active image: Architecture and engineering in the age of modeling* (pp. 77–97).

Johnson, G., Gross, M. D., Hong, J., & Do, E. Y. L. (2009). Computational support for sketching in design: A review. *Foundations and Trends® in Human–Computer Interaction, 2*(1), 1–93.

Jonson, B. (2002). Sketching now. *International Journal of Art & Design Education, 21*(3), 246–253.

Lewis, M. (2023). AIxArtist: A first-person tale of interacting with artificial intelligence to escape creative block. In *The 1st international workshop on explainable AI for the arts (XAIxArts), ACM creativity and cognition (C&C) 2023*. Online, 6 pages. https://arxiv.org/abs/2308.11424

Romic, B. (2022). Negotiating anthropomorphism in the Ai-Da robot. *International Journal of Social Robotics, 14*(10), 2083–2093.

Shafiei, S. B., & Esfahani, E. T. (2014, August). Aligning brain activity and sketch in multi-modal CAD interface. In *International Design Engineering Technical Conferences and Computers and Information in Engineering Conference* (Vol. 46285, p. V01AT02A096). American Society of Mechanical Engineers.

Sturdee, M., Hardy, J., Dunn, N., & Alexander, J. (2015, November). A public ideation of shape-changing applications. In *Proceedings of the 2015 international conference on interactive tabletops & surfaces* (pp. 219–228).

Sturdee, M., Lindley, J., Harrison, R., & Kluth, T. (2020, April). The seven year glitch: Unpacking beauty and despair in malfunction. In *Extended abstracts of the 2020 CHI conference on human factors in computing systems* (pp. 1–11).

Thomas, V., Brueggemann, M. J., & Feldman, D. (2015, September). I am more than the sum of my parts: An e-waste design fiction. In *EnviroInfo and ICT for sustainability 2015* (pp. 57–65). Atlantis Press.

Tresset, P., & Leymarie, F. F. (2013). Portrait drawing by Paul the robot. *Computers & Graphics, 37*(5), 348–363.

Williford, B., Runyon, M., Malla, A. H., Li, W., Linsey, J., & Hammond, T. (2017, October). Zensketch: A sketch-based game for improving line work. In *Extended abstracts publication of the annual symposium on computer-human interaction in play* (pp. 591–598).

Websites

W1 The home of Midjourney on the web – www.midjourney.com/
W2 The home of ChatGPT on the web – www.chat.openai.com/
W3 3D sketching pen – www.uk.the3doodler.com
W4 AxiDraw writing and Drawing machine – https://axidraw.com/
W5 The most recent version of DALLE on the web – www.openai.com/dall-e-3
W6 Milan Art Institute. 2021. Overcoming Artist's Block: 8 Ways for Finding Your Artsy Groove Again – www.milanartinstitute.com/blog/overcomingartist-s-block-8-ways-for-finding-your-artsy-grooveagain
W7 A4 Betabook – whiteboard notebook for on-the-go – https://www.kickstarter.com/projects/betabook/betabook-the-portable-whiteboard-for-the-digital-a

W8 A3 Whiteboard static sheets for indoor walls – https://www.amazon.co.uk/Magic-Whiteboard-Sheets-White-Erasable/dp/B019HHE4KS/ref=sr_1_7?crid=3H0U1SYX3GHSS&keywords=whiteboard+static+sheets&qid=1698425203&sprefix=whiteboard+static+sheets%2Caps%2C70&sr=8-7

W9 STAEDTLER Lumocolor Whiteboard Marker Bullet Tip - Assorted Colours – https://www.amazon.co.uk/STAEDTLER-351-WP4-Lumocolour-Multicolour/dp/B000J68GWY

W10 The Future of HCI – www.bootcamp.uxdesign.cc/the-future-of-hci-unveiling-a-world-beyond-our-wildest-dreams-5f46fa708367

Further Reading

Dunne, A., & Raby, F. (2013). *Speculative everything: Design, fiction, and social dreaming*. MIT Press.

Edmonds, E. A., Weakley, A., Candy, L., Fell, M., Knott, R., & Pauletto, S. (2005). The studio as laboratory: Combining creative practice and digital technology research. *International Journal of Human-Computer Studies, 63*(4–5), 452–481.

Kirman, B., Lindley, J., Blythe, M., Coulton, P., Lawson, S., Linehan, C., Maxwell, D., O'Hara, D., Sturdee, M., & Thomas, V. (2018). Playful research fiction: A fictional conference. In *Funology 2: From usability to enjoyment* (pp. 157–173).

Lewis, M., Lengyel, D., Sturdee, M., Toselli, M., Miers, J., Owen, V., Urban Davis, J., Gaudl, S., Xiao, L., Troisi, A., Priego, E., Mehnaz Huq, R., Snooks, K., Turmo Vidal, L., Blevis, E., Privato, N., Piedade, P., Claisse, C., Ford, C., Bryan-Kinns, N., Henriques, A., Severes, B., Kaipainen, K., Palosaari Eladhari, M., Grek, A., McMurchy, G., LC, R., Nabil, S., Jardine, J., Collins, R., Vlasov, A. V., Knight, Y., Cremaschi, M., Carderelli-Gronau, S., Núñez-Pacheco, C., Reyes-Cruz, G., & Rivière, J.-P. (2024). Travelling arts x HCI sketchbook: Exploring the intersection between artistic expression and human-computer interaction. In *Extended Abstracts of the CHI Conference on Human Factors in Computing Systems* (CHI EA '24), May 11–16, 2024, Honolulu, HI, USA. ACM, New York, NY, USA, 13 pages.

Chapter 14
Additional Resources and Community

14.1 Welcome to the World of Sketching in HCI!

Although you have now finished the book, your journey will continue. We hope that you have found your feet, sketched them, and are ready to put your skills to good use in your work and for fun. This final chapter looks at resources, personal style, continued practice via observational sketching, and inspiration. We hope you find yours!

14.2 Website and Social Media

The Sketching in HCI website (www.SketchingHCI.com) is the digital home of Sketching in HCI, populated by almost ten years (and counting!) of our research and teaching, new resources we think you might like, events, our own sketches, and of course, the Sketching in HCI course! We are constantly thinking of new ways to get people involved in sketching and would love to hear from you if you've found the book helpful, or simply want to share your sketches with us!

On social media, we use the hashtag #sketchingHCI and are currently active on several platforms; this is the quickest way to share with us, though feel free to reach out via our institutional emails. Social media is also a wonderful way to build community via different initiatives, or skill sets. For example, we connect with sketchnoters from all over the world and share our work, and Makayla regularly takes part in drawing challenges on "*X*" such as Today's Doodle (#todaysdoodle) a 365 daily sketching challenge, Inktober (#inktober) a series of prompts throughout October that encourage you to sketch with ink, and even Botober (#botober) a hilarious AI-generated prompt list for the same month. We have also mentioned the online (and sometimes in person, in London) fortnightly drink and draw that we like to

© The Author(s), under exclusive license to Springer Nature Switzerland AG 2024
M. Lewis, M. Sturdee, *Sketching in Human-Computer Interaction*,
https://doi.org/10.1007/978-3-031-50136-4_14

join—the Gosh Comics and *Broken Frontier* Drink and Draw #GOSHBFDD (Fig. 14.2)—and there are others out there!

Sharing your sketches in this way not only builds community with like-minded people but can cross over into making new connections for research and work too (see networking in Chap. 11). Starting a website or social media account such as *Instagram* or *Flickr* can help you keep all your work in one place—this can be kept private (or parts of it can) as you prefer and help save work if you lose the originals! The platform *X* is a fun way to connect but compresses images. You may have seen that Makayla often watermarks or adds copyright statements to her work; having these embedded helps keep the attribution obvious even if they are shared (Fig. 14.1)

You can also find a real-life sketching buddy, or even several sketching buddies! A sketching buddy should be someone who is happy to go on trips with you to galleries; sit in coffee shops, even stay at home; and sketch together with you. When we (Miriam and Makayla) are feeling demoralised or uninspired, having each other really helps. Mutual creativity and productivity = great sketches! For example, Figs. 14.3 and 14.4 where we sketch with our partners.

If you don't have an active sketching buddy, pets are also an endless source of joy and sketch inspiration; you may have spotted our pets in the book along the way… (Figs. 14.1 and 14.2).

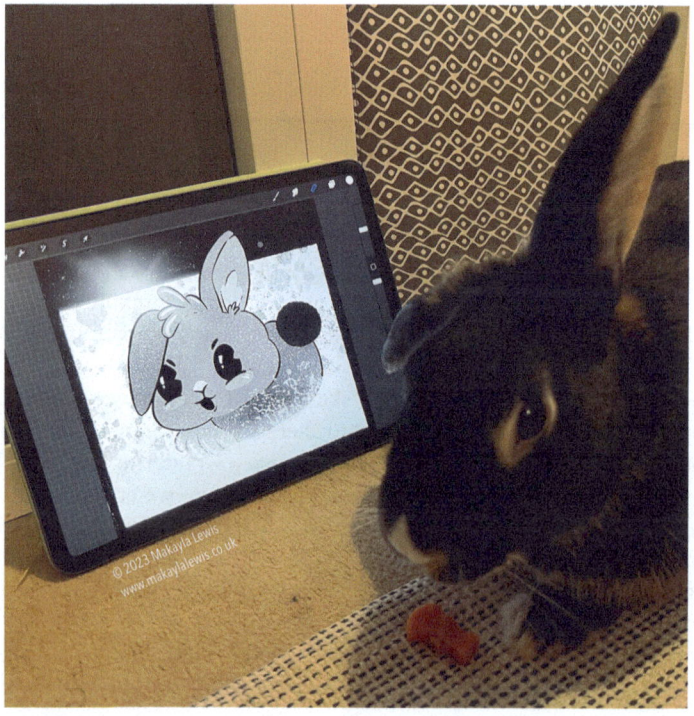

Fig. 14.1 Chinese New Year "Year of the Rabbit" next to 1-year-old rabbit "Umbriel". *Procreate* App on *Apple iPad Pro* using *Apple Pencil*. Makayla Lewis, 2023

Fig. 14.2 Tiny cat is unimpressed with Miriam's depiction of his recurring lunch… #GOSHBFFD prompt "puke banter". Photograph of sketchbook and cat. Miriam Sturdee, 2022

There are so many fabulous communities of sketchers out there—join them! And if real-life sketching community is your thing, join a local sketching group, or an in-person drink and draw? If there isn't one, why not start one?

14.3 Miriam's Favourite Personal Style

Personal style is an evolution. I spent so many years trying to force myself into a mould—with the desperate goal to be "realistic" that I lost sight of what was important! If we were all photorealistic sketchers, then the world would be a very boring place. Individual personal styles are what makes things exciting.

My favourite personal style, and a great "default" when I want to enjoy sketching and not stress out, is definitely my colour layering sketching technique, used with

Fig. 14.3 Sketch for Makayla's birthday card. *Procreate* App on *Apple iPad Pro* using *Apple Pencil*. *Procreate* App on *Apple iPad Pro* using *Apple Pencil*. James E O'Donnell, 2023

Fig. 14.4 Sketch of #GOSHBFDD prompt "wholesome violence". Fineliner pen, pencil and marker on paper. Paul Whaley, 2021

14.4 Makayla's Personal Style

either detailed outlining or a scribbly style and overlay. Sometimes I try to make things neat and tidy, but the result is much better when I stop thinking and *START SKETCHING*.

This style focuses on outlining and doodling into shapes and around colours. This technique can be more specific (Fig. 14.5), or I can keep it loose and use patterns and fills (Fig. 14.6), with a resulting more abstract image. I've also used this with paint and ink (Fig. 14.7), which can be quite unpredictable!

If you want to try it for yourself, the first thing to do is get some dual-tip markers like *Copic*. Having the chisel and fine nib means you can use the same colour in different ways. You also want to have a range of colours that either contrast each other, allowing for a fun palette, or of the same colour and different hues so you can build gradients. The key is to always have a colour that is light enough to sketch with, and almost invisible in the final result. Each layer adds more confidence of line, until you feel you have a scene you want to work into with your fineliner.

At this stage, you can either take the approach of outlining *EVERYTHING*, each pen stroke and colour, and build a fully complex and abstract image, including pattern and hatching if you wish, or draw *PAST* the coloured lines and areas to overlay a more naturalistic line view, or a combination of both! This technique works well with ink and paint too, for a less controlled approach to the initial layers.

Fig. 14.5 Beagle, Quick-fire colour layer and scribble sketch. Fineliner pen and marker on paper. Miriam Sturdee, 2012

Fig. 14.6 Umbrellas. Inspired by living in Japan. Fineliner pen and marker on paper. Miriam Sturdee, 2011

14.4 Makayla's Personal Style

Fig. 14.7 Japanese drawing ink sketch for a magazine sketch feature. Fineliner pen and ink on paper. Miriam Sturdee, 2023

Practical Application Tips

- Try random colour palettes that have good colour contrast; you can build a scene with two or three colours to add a new spin on your work.
- Or, try greyscale for the majority of a scene, and highlight one object or person in colour (as in Fig. 14.6).

- Make sure if you are going to overlay colours that they do not bleed into each other, especially when you are creating outlines or want to add more colour later—always test your pens and paper!
- Don't try to force the image to look realistic; often it works better if it is rough and wonky.
- Don't overthink things; this is a style that is loose and rewards those who are happy to play around with ideas.
- Do have fun, and try out a few variations!

14.4 Makayla's Personal Style

"Practice does not make perfect; it creates another experience"—*Makayla Lewis, 2023*

During a late evening co-writing session for this book, Miriam asked me, "Can you wang out a section on your favourite personal style then return to Chapter 8, please?". I have often been asked this question, and my answer has not changed. I am sorry to say, Miriam; I do not have a static style; I am an experimental sketcher and illustrator, e.g. Figs. 14.8, 14.9, 14.10, 14.11, 14.12, 14.13, 14.14, 14.15, and 14.16. Over the years, I've tried lots of new sketching techniques and tools. I enjoy searching out possibilities through other artists (all the books in this book I own; yes, before you ask, my bookcase shelves are bowing) and tools (I have hundreds of markers, and I am not ashamed to say I love each one).

Experimental, sketching allows me to do the following:

- Push the boundaries of my creative process and discover new ways to sketch.
- Trying new techniques and tools has enhanced my skills and experience in sketching, thus making me a more versatile sketcher.
- Allow me to explore different styles freely without the guilt of not being known for one style.
- It has allowed me to solve problems in different contexts and for different needs, as I often find one style only fits some.
- Trying new tools and techniques inspires me to come up with and explore ideas in varied ways.
- Sharing my experimentation often provokes valuable discussions and opportunities for networking.

Oh, I forgot to mention; I enjoy being an experimental sketcher. I do not think I will ever change; if you look at my research papers, they show this experimentation now and then, e.g. virtual reality, autoethnography, money management, accessibility, change management, artificial intelligence, design thinking, cybersecurity, and sketching. Trying new things is ingrained in my being; some people find experimentation scary, but I embrace it.

14.4 Makayla's Personal Style

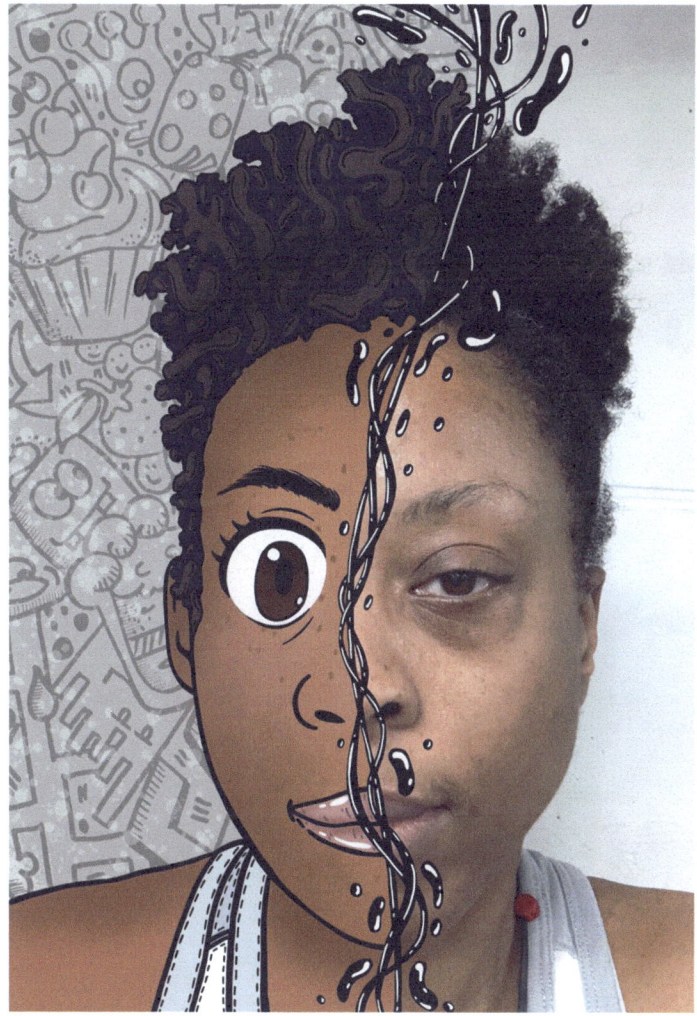

Fig. 14.8 Makayla's response to ToonMe challenge on *Instagram*. *Procreate* App on *Apple iPad Pro* using *Apple Pencil*. Makayla Lewis, 2020

Practical Application Tips

- Embrace the unknown; if you don't, you won't experiment.
- Try many diverse materials; even if they scare you, you will experience something new.
- An experiment in different locations: people, objects, and sounds (the hustle and bustle of everyday life) are everywhere we go; this may inspire you to create differently.
- Mix and match your styles to create new styles, which can be bad but sometimes good.

Fig. 14.9 Two hours in a Makayla's sketchbook. Fineliner pen, watercolour, colour pencil and marker. Makayla Lewis, 2022

Fig. 14.10 Sketchnotes from visit to Tate Modern and Tate Britain, London. Fineliner pen and marker on paper. Makayla Lewis, 2019

Fig. 14.11 Sci-fi landscapes, fineliner pen and paint marker. Makayla Lewis, 2019

Fig. 14.12 "What will happen if I say nothing?". *Procreate* App on *Apple iPad Pro* using *Apple Pencil*. Makayla Lewis, 2021

Fig. 14.13 Snap! Fineliner pen, white gel pen, and marker on paper. Makayla Lewis, 2019

- Give yourself time constraints and throw away the eraser, as this will hold you back.
- Limit your colour palette as it will make you think outside the box, e.g. if an object is dark green, but you only have blue, how will you convey it, maybe a feeling?
- Reflect on your sketches, and further experiment on the ones that work; it is okay to abandon others.
- Have fun (Figs. 14.17 and 14.18).

14.5 Observational Sketching

Fig. 14.14 Sketchnotes from Research.Thing VR and Research. Fineliner pen and marker on paper. Makayla Lewis, 2023

Fig. 14.15 Sketchnotes from *Sketchnote Hangout* digital visit to Mauritshuis, NL. Fineliner pen and marker on paper. Makayla Lewis, 2020

Fig. 14.16 Doodling away in a school of mathematics and computer science staff meeting. *Procreate* App on *Apple iPad Pro* using *Apple Pencil*. Makayla Lewis, 2023

Fig. 14.17 "TLE". *Procreate* App on *Apple iPad Pro* using *Apple Pencil* from (Lewis, 2024)

Fig. 14.18 "Welcome to Roundville" a response to Mauro Toselli Phrenville sketching game (W7). Fineliner pen on paper. Makayla Lewis, 2021

14.5 Observational Sketching

A question we are often asked is, "Now I have the basics, how can I improve EVEN MORE?" The obvious answer is keep practising the basics and elaborate upon them, but there is another part of sketching practice that will see you progress in leaps and bounds and bring you close to achieving the thing we advocate most—**personal style**!

The answer is observational sketching. Which is exactly what it sounds like. Look at your environment, wherever you are, and sketch it, be it people, places, buildings, cats, dogs, frogs... Visit interesting places and sketch items or scenes of interest, or visit boring places and sketch your surroundings to pass the time (Figs. 14.19, 14.20 and 14.21). You might even consider your sketchbook to be a piece of "reportage", documenting an event or moment in someone's life.

Your own home can also be a source of inspiration if you cannot go out. Even the everyday and mundane can be interesting to sketch, and it helps us notice new things about our environment (Fig. 14.22). How does washing hang on the line? How does the fold in the blanket fall when it has been thrown onto the sofa? Sketching is more than just looking; it allows for a deep interrogation of the visual world.

One extra piece of guidance that is also useful is to go to life-drawing classes if you wish to improve your understanding of drawing people. Physiology can be complex, and understanding the underlying musculature can improve your sketches of people. If there aren't any classes in your area however, or cost is prohibitive, draw your friends and family (clothes on!), doing activities they enjoy. Pro tip—if your friends like video games, they are a great starting point for sketching figures from life as they tend to stay fairly still... Alternatively, sit with napping relatives during the holidays and sketch them! (Fig. 14.23)

Get comfortable! Make use of what is available (e.g. benches, walls, chairs, even the floor if it is clean!). Sometimes you may have to stay standing, in which case

Fig. 14.19 Observational sketch in a coffee shop. *Photoshop* on *Microsoft Surface Pro* using *Microsoft Surface Pen*. Makayla Lewis, 2016

14.7 Books and Media

Fig. 14.20 Observational sketch in a Swedish coffee shop. Fineliner pen and marker on paper. Miriam Sturdee, 2023

Fig. 14.21 Observational sketch from Tate Britain Observation Deck. Photograph and *Procreate* App on *Apple iPad Pro* using *Apple Pencil*. Makayla Lewis, 2019

Fig. 14.22 Observational sketch of the kitchen sink. Fineliner pen and marker on paper. Miriam Sturdee, 2021

14.7 Books and Media

Fig 14.23 Looking outwards, looking inwards, rooftops in Rome, and a sleeping person. Fineliner pen and marker on paper. Miriam Sturdee, 2018

Fig. 14.24 Makayla sketching on location. Photograph. Makayla Lewis, 2016

having a small, light sketchbook and easy-flow pen helps so you can quickly get the basics down without taking a long time and getting arm ache. It also helps if you wear comfortable clothing, comfortable shoes, and a non-cumbersome bag or backpack (Fig. 14.24). If you sketch standing up and have a semblance of the scene you are trying to represent, you can then retire to a place where you have a table and chair and work on it there, or go home and add details and colour later.

It will never be possible to sketch everything you see on location—we encourage you to focus in on the details (context, people, and technology) omitting or limiting backgrounds to depict what is of most interest. Think also how big something is in relation to its surroundings ("sighting"), although "artistic licence" comes into play here—realism is NOT necessary and your own style may preclude perspective and logic! For example, the rooftops in Fig. 14.23 are juxtaposed without adherence to accurate angles or depth, but the result is still a great sketch. Figures 14.25, 14.26, 14.27 and 14.28 show different approaches to sketching on location, some focused on objects and details, some providing a full overview of the place and its exhibits. You should choose your own approach by the inspiration you have on the day and the materials you have with you!

14.8 The Sketching in HCI Manifesto

Fig. 14.25 Makayla visits the sculpture in Britain 1600–1950 at the V&A Museum, London. Fineliner pen and marker on paper from (Lewis et al. 2023b)

These "quick and loose" sketches are the backbone of drawing on location. Some people even find they can use other materials to make sketches of scenes, such as watercolours and brush or watercolour pencils and water pen. These kinds of materials have a flow which encourages looseness. If this approach does not appeal, use light coloured, chunky markers, and sketch the "bones" of the scene, and then sketch in finer details in fineliner as a reference for later. You can also reduce the detail and loose nature of the sketch and refine the imagery into simple scenes for a storyboard.

Fig. 14.26 Sketchnotes from Cartoon Museum, London. Fineliner pen and marker on paper. Makayla Lewis, 2023

Observational sketching is not a quick task or activity to be completed. It is a journey. A sketchbook is ideal to show your skills develop over time and encourages you to think "one page at a time". Don't like the result? Move on and keep sketching. Much like learning a new language, continued practice will result in improvement and development of your personal style. Break out of the confines of realism, and look at the world via the lens of your pen!

14.8 The Sketching in HCI Manifesto

Fig. 14.27 Sketchnotes the British Museum, London. Fineliner pen and marker on paper. Makayla Lewis, 2023

Practical Application Tips

- Get yourself a small A5 or A6 sketchbook that you can keep in a handbag or pocket. Make sure it is easy to hold open on a single page; spiral binding can help with this, or travel notebooks designed especially for the job.
- Leave super fine pens at home; get something that encourages flow and economy of line rather than fine grained detail.
- Don't start on the first page—a new sketchbook can be intimidating. Instead, pick a middle page at random.
- Pick an "on-the-go" sketching set including a couple of colours and a grey marker. Yes, we know it is hard to limit your pen selection!
- If you make a mistake, just keep going and draw over it. Starting in lighter pen colours helps.
- People are always fascinated by people sketching in public, but don't feel obliged to share unless you are comfortable doing so.

14.6 Photographing and Scanning Your Sketches

You cannot always use digital, or may not wish to, so we have put together a quick guide to photographing and scanning your sketches.

First of all, make sure you are in a well-lit room! Avoid yellow light if possible, and try to find an angle where your hand and camera or camera phone are not causing a shadow. If you can't eliminate the shadow entirely, move back and take the

Fig. 14.28 Sketches from Kew Gardens London "Orchids: Celebrate the colour of Columbia" exhibition. Fineliner pen and marker on paper. Makayla Lewis, 2019

highest-resolution image you can with the image in the unshadowed part—this can then be cropped later digitally. Natural daylight is best, or a combination of light sources. You can get a small-angle poise lamp with a blue tone lightbulb to help illuminate the image.

Unless you want a shot "in context" with your sketchbook pages and source or to show your materials (useful for social media sharing), try to avoid clutter and objects on or around your sketched page. It is much more difficult to edit things out later on. You should also try to line your page up straight in the shot, rather than at an angle. You ideally want a high contrast, closely cropped shot, especially if you are not going to edit it later on the computer.

Another lesson we have learned is to take photos *in progress…* sometimes you might prefer an earlier version (see Fig. 14.29) or take a photo and lose the original. This can backfire when you are not explicitly taking the earlier image for later use, as the quality, light, and angle may not be ideal (we are guilty of this too!), so try to build process photos into your sketching practice. It is also fun to look back on how your features, layers, and layout developed.

Scanning is usually easier than taking a good photograph, but you may not always have access to a scanner, for example, at home or when travelling. In these cases, photography is your best bet. If you are lucky enough to have a scanner, then the first thing to consider is the resolution and output, and if you will be doing any post hoc editing. The standard for print scanning is 300dpi, and usually we go for a PNG output, though some journals and publishers like to use TIFFs as they do not lose quality (and support CMYK printing—four-colour plate printing). JPGs tend to be lossy so are best avoided if possible, though are very easy to share and do not have transparency issues. Make sure you also have the correct colour settings! RGB is best for screen output, but beware when printing RGB as the colours will never be as vibrant. CMYK images are usually larger, different formats (TIFF) and will display the on-screen colour closer to the final print colour. Generally though, unless you are creating a printed book or pamphlet in large quantities, RGB is fine, and many self-publishing places will simply print RGB files.

One word of warning however… some coloured markers we have used change colour when scanned or photographed! So if you have made a colour palette on paper, try photographing and/or scanning it to check the colours look how you expected.

Fig. 14.29 Miriam's sketch of digital entrepreneur Peter Kariuki for ACM Interactions magazine (Sturdee, 2024). Having photographed an earlier stage to show a friend, I later found I preferred the rougher, sketchier image without the additional detail and penwork. Fineliner pen and marker on Bristol board. Miriam Sturdee, 2023

14.7 Books and Media

We have already recommended books and websites in each chapter which you may find helpful or interesting, and we will also update the website as new materials and resources come out. However, we also thought it would be fun to each share our favourite sources of inspiration here.

14.7.1 Miriam's Inspiration

I've always been drawn to artists who have a sketchy, effortless style, yet make it an artform. During my MFA in Edinburgh, I was lucky enough to visit an exhibition at the Dean Gallery by an Edinburgh College of Art graduate called Charles Avery. He had created a fictional world, an island, a whole people and cornucopia of animals and rituals. There were loose sketches on brown paper using white and brown pencil and black ink; there were snapshots of interaction and villages, even a stuffed animal that does not exist. I was captivated that this world and these sketches conveyed so much meaning and history and desperately wished for the accompanying book—but it had sold out (Avery et al., 2008). Years later it was gifted to me, and simply flicking through its beautiful sketches and photographs of the exhibition transports me back to art college and continues to inspire me.

I couldn't limit my inspiration to only one resource, so I also want to mention Ronald Searle's *To the Kwai and Back: War Drawings 1939–1945* (Searle, 1986). Searle was a reportage illustrator who began sketching daily life after his capture and incarceration in prisoner of war camps in Japan during the Second World War. He smuggled his sketches from place to place, begging and bartering for sketching supplies, and this book collects these inspiring works and their story in one place. Searle reminds me of the value of sketching, as a passion, as a retreat, and as a gift.

Aside from printed books and mindful activities such as cycling, I find inspiration in working with Makayla—hence the creation of our book! As having someone on the same sketching page as yourself means you are never alone. Sketching in HCI has been a long long journey (Fig. 14.30).

14.7.2 Makayla's Inspiration

Unlike Miriam, I was not formally trained; I have always sketched, which I have done since I was a child—I did draw on walls. It was a hobby that remained in the shadows. When I was introduced to HCI during my undergraduate degree at Kingston University London, I immediately saw the connection between visual thinking and HCI; it was an "Aha" moment. I started sketching experiences, creating rudimentary visual notes (before they became known as sketchnotes), and summarising my ideas and findings through narrative sketching (storyboards).

Fig. 14.30 Self-portrait, A long time ago… Fineliner pen and marker on paper. Miriam Sturdee, 2023

The overlap between sketching and HCI took hold and never left—granted, it was on the periphery of my everyday life (for my eyes only). During my first postdoc at Royal Holloway, University of London, however, I leapt and firmly decided that my sketching practice and research practice are intertwined—my pen and paper should not be hidden on the page under my desk. I want to thank the international sketchnote community for showing me that hiding one's creative pursuit is a bad idea: one of my sketchnotes at the time was published in Mike Rodhe's sketchnote workbook—the moment I received the email asking will never be forgotten—thus heralding the launch of my new "open" sketching research era. Sketching began to permeate my research practice and everyday interactions with the world.

I started taking part in TodaysDoodle on social media (a challenge to sketch for 365 days), which I completed but have continued to upload to the hashtag occasionally, currently at No.917 (wow, I have shared almost 1,000 doodles/sketches/illustrations on *Flickr* (W3 and W4), *X* (W5), and *Instagram* (W6) over the last 6-ish years, e.g. Fig. 14.31). As a result of this excessive sharing, I was invited to sketch events, teach sketching, and do great freelance work (W2), which I have been doing ever since, and, of course, meeting Miriam on *X*, and well, everything else is history, well the present and the future.

So, back to the section heading, what inspires me… **museums, galleries, and installations** (exploring different artists, both online and in-person, with others, e.g. Lewis et al., 2022a), **co-creation** (sketching with others especially colleagues and

Fig. 14.31 TodaysDoodle No. 899 Self-portrait. *Procreate* App on *Apple iPad Pro* using *Apple Pencil*. Makayla Lewis, 2022

users), **teaching and learning** (seeing others explore and practise sketching, e.g. Lewis & Sturdee, 2022), **manga and comics** (offer different techniques and structures when visual storytelling), **animation** (different possibilities of character design and building engaging visual stories, e.g. Star Wars animations and classic Disney movies), and **video games** (the interactions between characters and the player, the digital worlds in which "we" reside, and the elaborate ways of telling stories, e.g. cut scenes).

I do not imagine putting down analogue or digital pencil and paper anytime soon.

14.8 The Sketching in HCI Manifesto

We want to finish by referring to the Sketching in HCI manifesto that summarises the book's key message. It was facilitated, analysed, and crafted by **Makayla Lewis**, **Miriam Sturdee**, **Thuong Hoang**, **Pranjal Jain**, **Mafalda Gambol**, **Katta Spiel**, **Ernesto Priego**, **Nicolai Marquardt**, **Marina Fernández Camporro**, **Jagoda**

Walny, **Joanna Foster**, and **Sheelagh Carpendale** over 5 years at ACM SIGCHI conference in the form of a recurring special interest group.

The creation of the manifesto was not an insular process; the **talented international ACM SIGCHI community (students, researchers, and practitioners)**, those who attended the many special interest groups over the years, were fundamental in its creation; thank you for sharing your experiences, needs, preferences, and pain points.

The manifesto was collected from four special interest groups (Sturdee et al., 2021; Lewis et al., 2018; Lewis et al. (2022a, b), and the community convened to consider inclusion and accessibility at the fifth special interest group in 2023 (Lewis et al., 2023b):

Role—Sketching in HCI can be called upon to explore and communicate everyday research.

Creator vs. viewer—Sketching in HCI, creators should consider the aims and the views of their sketch before and during creation to support sketch relevancy.

Publication—Sketching in HCI should be viewed as a valid dissemination of HCI research and practice.

Not alternative—Sketching in HCI should be viewed as something other than an alternative communication medium but should hold equal footing with different visual outputs.

Training and practice—Sketching in HCI should not be restricted to arts and creatives and should be incorporated into HCI education.

Funding and impact—Sketching in HCI can be used to bridge the gap between academics and practice by easily, concisely, and quickly disseminating ideas and learnings.

Inclusive—Sketching in HCI should represent our world, ethnicities, genders, sexuality, religion, geographies, and cultures.

Accessibility—Sketching in HCI should be accessible to all; creators should include alternative methods for engagement, e.g. AltText, AltNarrative, sound, language, and touch.

Technology—Sketching in HCI should not rely on technology; creators should consider their audience's skills, needs, and preferences.

14.9 Invitation

Not only do we invite you to engage with Sketching in HCI but also to reach out to us with questions, to share your sketches, and to discuss past, current, and future Sketching in HCI research and practice!

We look forward to hearing from you!

14.10 Hands-On Activities

Activity 14.1: Maintaining Your Practice (Individual)
Learning objective—To keep sketching and learn from observing the world
Time—5 minutes to infinity+
Materials—By now we hope you have found your preferred materials, perhaps a sketchbook and a fineliner, maybe even watercolours for the adventurous!
Procedure:

- When you are out and about, at a bus stop, at a coffee shop, even in your kitchen, take out your sketchbook and materials, and draw from real life—we call this observational drawing.
- Try to sketch in different places and situations, if you have already made lots of sketches of buildings, try adding people.
- If you have drawn a lot of scenes and people, try sketching some items as a still life.
- It helps to have a range of imagery to practise on; you will refine your technique and further develop your personal style.
- Try life drawing for a better understanding of human anatomy and to have a change of pace.
- Quick sketches—of about 30 seconds—of moving objects are a great way of discovering economy of line.
- Just keep sketching!

We are so pleased that you have joined us on this journey through Sketching in HCI—we hope you have enjoyed reading and interacting with this book as much as we have enjoyed reading it. We also hope that you are able to continue your practice and utilise your newfound skills at work and play!

References

Books and Papers

Avery, C., Bourriaud, N., & Morton, T. (2008). The Islanders: An introduction [published on the Occasion of the Exhibition Held at the Parasol Unit Foundation for Contemporary Art, London, 10 September-8 November, 2008; Scottish National Gallery of Modern Art, Edinburgh, 29 November, 2008-15 February, 2009; Museum Boijmans Van Beuningen, Rotterdam, 28 February-24 May, 2009]. Koenig Books.

Lewis, M. (2024). Looking back, moving forward: A first-person perspective of how past artificial intelligence encounters shape today's creative practice. In *Proceedings of Explainable AI for the Arts Workshop 2024 (XAIxArts 2024) arXiv:2406.14485* (p. 6). ACM, New York, NY, USA.

Lewis, M., & Sturdee, M. (2022). Curricula design & pedagogy for sketching within HCI & UX education. *Frontiers in Computer Science, 4*, 826445.

Lewis, M., Sturdee, M., Marquardt, N., & Hoang, T. (2018). SketCHI: Hands-on special interest group on sketching in HCI. In *Extended abstracts of the 2018 CHI conference on human factors in computing systems (CHI EA '18)*. Association for Computing Machinery, New York, NY, USA, Paper SIG09, 1–4.

Lewis, M., Toselli, M., Baker, R., Rédei, J., & Ohlenschlager, C. E. (2022a, June). Portraying what is in front of you: Virtual tours and online whiteboards to facilitate art practice during the COVID-19 pandemic. In *Proceedings of the 14th conference on creativity and cognition* (pp. 350–363).

Lewis, M., Sturdee, M., Miers, J., Davis, J. U., & Hoang, T. (2022b). Exploring AltNarrative in HCI imagery and comics. AltChi paper to 2022 CHI conference on human factors in computing systems. Association for Computing Machinery, New York, NY, USA, Article 164.

Lewis, M., Sturdee, M., Gamboa, M., & Lengyel, D. (2023a, April). Doodle away: An autoethnographic exploration of doodling as a strategy for self-control strength in online spaces. In *Extended abstracts of the 2023 CHI conference on human factors in computing systems* (pp. 1–13).

Lewis, M., Sturdee, M., Hoang, T., Gamboa, M., & Jain, P. (2023b). SketCHI 5.0 diversity & accessibility at the core of sketching in HCI. In *CHI conference on human factors in computing systems extended abstracts (CHI '23 extended abstracts)*, April 23–28, 2023, Hamburg, Germany. ACM, New York, NY, USA.

Searle, R. (1986). *To the Kwai and back: War drawings 1939–1945*. Souvenir Press Ltd..

Sturdee, M., Lewis, M., Spiel, K., Priego, E., Camporro, M. F., & Hoang, T. (2021). SketCHI 4.0: Hands-on special interest group on remote sketching in HCI. In *Extended abstracts of the 2021 CHI conference on human factors in computing systems. association for computing machinery, New York, NY, USA*, Article 164, 1–4.

Sturdee, M. (2024). Keeping things real with Peter Kariuki. *Interactions, 31*(1), 10–11. ACM.

Websites

W1 Join us! – www.SketchingHCI.com
W2 Makayla's website (profile, blog, portfolio, social media and contact – www.makayla-lewis.co.uk
W3 Makayla TodaysDoodle gallery on Flickr – https://www.flickr.com/photos/makaylalewis/sets/72157648616675866/
W4 Makayla Sketchnotes gallery on Flickr – https://www.flickr.com/photos/makaylalewis/sets/72157633090981769
W5 Makayla X profile – https://twitter.com/maccymacx
W6 Makayla Instagram profile – https://www.flickr.com/photos/makaylalewis/sets/72157648616675866/
W7 Mauro Toselli creator of Phrenville Game – https://www.maurotoselli.com/

Further Reading

We recommend you explore the wonderful world of sketching books whilst out and about; visit art stores, galleries, and secondhand bookshops; find your favourites; and treasure them.

SPRINGER NATURE

GPSR Compliance

The European Union's (EU) General Product Safety Regulation (GPSR) is a set of rules that requires consumer products to be safe and our obligations to ensure this.

If you have any concerns about our products, you can contact us on ProductSafety@springernature.com

In case Publisher is established outside the EU, the EU authorized representative is:

Springer Nature Customer Service Center GmbH
Europaplatz 3
69115 Heidelberg, Germany

The manufacturer's authorised representative in the EU is Springer Nature Customer Service Centre GmbH, Europaplatz 3, 69115 Heidelberg, Germany. If you have any concerns regarding our products, please contact ProductSafety@springernature.com

Printed and bound by CPI Group (UK) Ltd, Croydon, CR0 4YY

02/02/2026

02046252-0012